Lab Manual

FOURTH EDITION

physics

bju press

Greenville, South Carolina

The writers and publisher have made every effort to ensure that the laboratory exercises in this publication are safe when conducted according to the instructions provided. We assume no responsibility for any injury or damage caused or sustained while performing activities in this book. Conventional and homeschool teachers, parents, and guardians should closely supervise students who perform the exercises in this manual. More specific safety information is contained in the *Physics* Teacher Lab Manual Fourth Edition, published by BJU Press. Therefore, it is highly recommended that the Teacher Lab Manual be used in conjunction with this manual.

NOTE: The fact that materials produced by other publishers may be referred to in this volume does not constitute an endorsement of the content or theological position of materials produced by such publishers. Any references and ancillary materials are listed as an aid to the student or the teacher and in an attempt to maintain the accepted academic standards of the publishing industry.

PHYSICS Student Lab Manual
Fourth Edition

Writers
David M. Quigley, MEd
Bob Hill, EdD

Consultants
Christopher D. Coyle
Kelly Driskell
Hannah Lovegrove

Biblical Worldview
Renton Rathbun, PhD
Bryan Smith, PhD
Tyler Trometer, MDiv

Academic Integrity
Jeff Heath, EdD

Academic Oversight
Rachel Santopietro, MEd

Editor
Rick Vasso, MDiv

Designers
Garison Plourde
Emily Rush
Brenna Short

Illustrators
Ciara Chafin
Briseydi Rouse
Lovett Design

Page Layout
Carrie Walker

Permissions
Sylvia Gass
Lily Kielmeyer
Hannah Labadorf

Project Coordinator
Heather Chisholm

Digital Content Management
Peggy Hargis

Photo credits appear on pages 346–47.

The text for this book is set in Adobe Minion Pro, Adobe Myriad Pro, Arial MT, Helvetica, Minion Math by Typoma GmbH, Playfair by Claus Eggers Sørensen, Raleway, Segoe, and STIX fonts.

The cover shows a close-up of a powder explosion captured with high-speed photography against a black backdrop.

ISBN 978-1-62856-964-3

15 14 13 12 11 10 9 8 7 6 5 4 3 2

CONTENTS

CONTENTS

21 Refraction

22 Wave Optics

Unit 6 Modern Physics

23 Relativity

24 Quantum Physics

25 Nuclear Physics

Appendixes

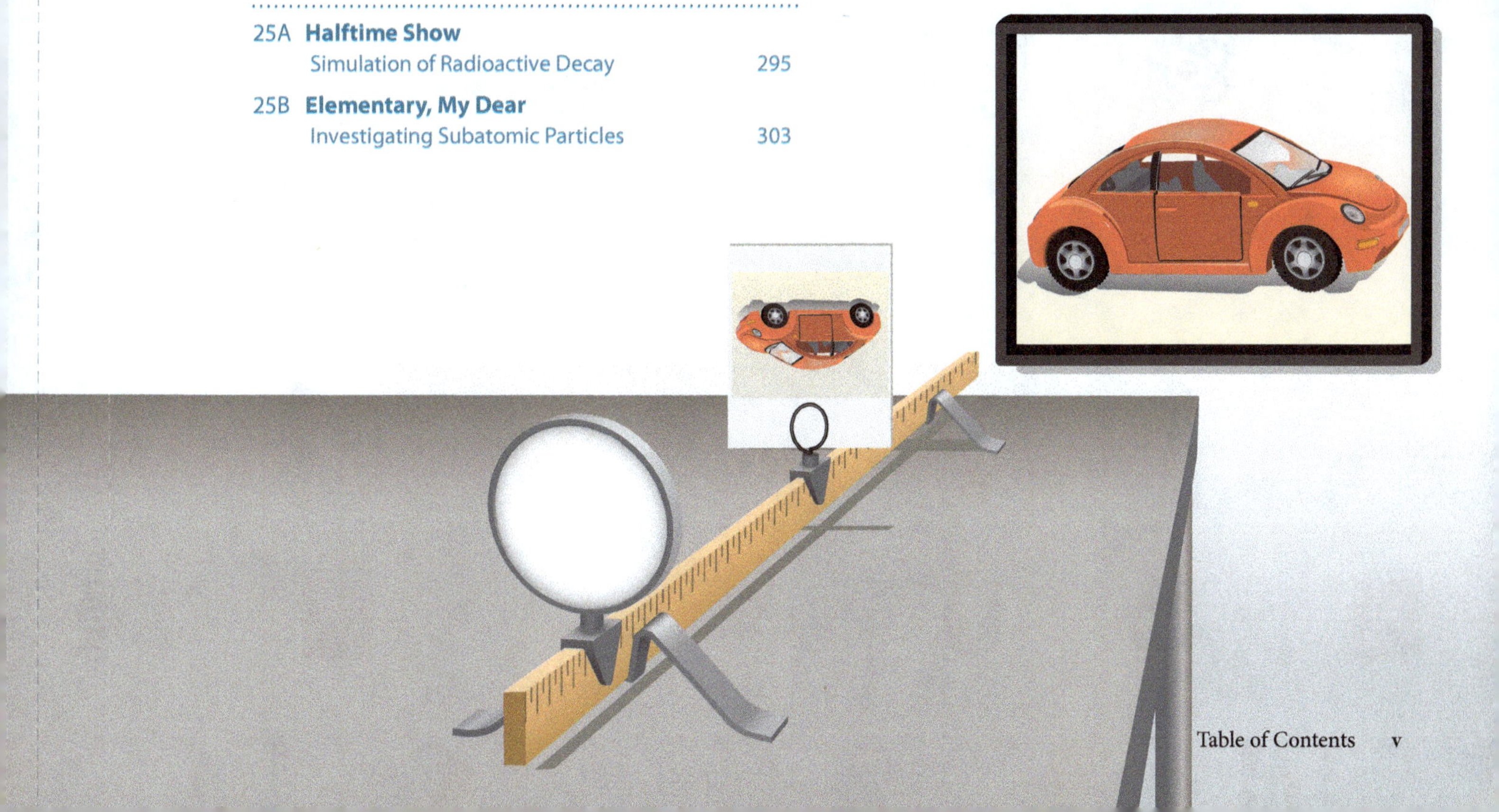

Welcome to the Laboratory

Everything is physics! Your diaphragm moves to create a partial vacuum that draws air into your lungs. Atoms and molecules collide to produce a chemical reaction. A football player applies a force to pass the ball. A cheerleader rotates to do a perfect backflip. A photographer adjusts the aperture of her camera to get the perfect picture. A musician alters vibrational patterns to produce beautiful music.

Much of what we know about physics was learned in the laboratory. The interesting thing is that a physics laboratory can be anywhere. Yes, some physics discoveries are made in a typical science laboratory, but others are made in ordinary, everyday places. Physics experiments can be done on a racetrack, in the ocean, on the moon, and in a music hall, just to name a few places. The real point is that since everything is physics, we can and should study physics everywhere.

THINKING LIKE A PHYSICIST

"Who, me?" Yes, you!

"But I'm just a high-school physics student!"

But you can be a student physicist! Even if you don't see a science career in your future, you can develop skills this year to help you work like a physicist at your own level. As you work through this lab manual, you will learn to think like a scientist—being safe in the laboratory, making predictions, collecting data, and testing your ideas. In some of the lab activities, you will have to think about how to solve a problem on your own, without any procedures spelled out in the lab manual. These inquiry lab activities will have the word *inquiring* in the subtitle. You will also do the work of an engineer in STEM activities, in which you will design and build an object and then test its effectiveness at solving a specific problem. STEM lab activities all have subtitles that begin with the word *building* or *designing*. You'll learn how to use equipment to collect data and how to tell whether the data that you've collected is good. The equipment for these activities can come in different shapes and sizes, both high-tech and low-tech. You'll learn to use masses, stopwatches, meter sticks, and dynamics carts. But you'll also use smartphone apps, computer programs, the internet, and possibly probeware.

Above all, learn to do science within the framework of a biblical worldview. Everyone has a worldview. It is the lens through which we view everything. This worldview doesn't just affect what a person believes about God, Jesus, and salvation—it also affects how he understands science, makes sense of history, and even how he interacts with other people. In science, worldview greatly affects what a scientist decides to study and how he interprets data.

As a student scientist, keep your worldview in mind. When you do an experiment, think about how it relates to a biblical worldview. A Christian student physicist interprets the world as God has already interpreted it in His Word. Christians should see science as an important gift of God, given to humans so that they can obey the Creation Mandate (Gen. 1:28) and the two great commandments (Matt. 22:37–40). When Christians do science, they bring glory to God and help their fellow humans to live in this fallen world. Therefore, some questions will specifically ask you to use your worldview to apply the information from your lab activity to a real-world problem. You will be challenged to consciously think like a Christian as you're doing science.

A Christian shouldn't see science as a pathway to truth. Rather, he should look to the Bible, the one source of absolute truth, to establish the guidelines and proper role of science in human life. Scientists working from a biblical worldview should evaluate existing scientific models in light of Scripture and challenge or revise them when they contradict Scripture. And as they create models of their own, they must test them for agreement with the revealed truths in the Bible. When we act in this way, we glorify God by using science as He intended.

So... let's get into the laboratory!

KEYS TO LABORATORY SUCCESS

Lab activities can be both interesting and fun. But your laboratory experience largely depends on you. If you take these activities seriously, keep your brain engaged, and try to learn as much as possible, you'll come away from the laboratory feeling that your time has been well spent. On the other hand, if you treat lab activities carelessly, you probably won't learn much nor will you have a genuine sense of accomplishment. The following guidelines will help you get the most from lab activities. And they'll also help you experience the true pleasure of a brain at work.

1. Read through the entire lab procedure before class. You'll know what to expect, and you'll be far less likely to make mistakes or forget a crucial step.
2. Review the parts of your textbook that apply to the lab activity. Your teacher may give you specific review guidelines.
3. Come to class well prepared. Bring your textbook, calculator, paper, and pencil. You won't necessarily need all these for every lab activity, but you'll regret it if they're sitting in your locker on a day when you *do* need them!
4. Begin the lab activity only when instructed to do so. Read each lab procedure step carefully so that you don't make mistakes. Procedure steps are identified by lettered bullets. If you're ever in doubt about what to do, *ask*; don't guess.
5. Lab activity questions are mixed in with the procedure steps. The questions are identified by numbers.
6. Most of the lab activities require you to record data in tables. For some of the lab activities you will be provided with data tables, and these will typically be located at the end of the activity. In other lab activities you will collect data electronically and then enter that data into a spreadsheet. Your teacher will tell you how to turn in the data for those lab activities.

7. Record your results from measurements and calculations carefully and accurately. Remember to record using appropriate significant figures.
8. Most lab activity questions ask you to explain your answer. Don't view this requirement as pointless busywork. Explaining something makes you think more deeply about your answer. It also helps you connect principles from the textbook, class discussion, and the real world to what you're doing in the laboratory.
9. Keep your laboratory area tidy. At the end of class, put away your equipment and dispose of trash according to your teacher's instructions.
10. While socializing may seem like a lot more fun than the lab activity, leave that behavior outside the laboratory door. You'll be more focused, get better results, and make far fewer mistakes.

STAYING SAFE

Nothing ruins a day quite like getting hurt in the laboratory. While the lab activities in this manual have been carefully designed to be safe, no lab activity is completely free of potential danger. For this reason, you should maintain constant vigilance when you're working through one.

Appendix A covers laboratory safety and basic first aid. Review that appendix before doing any of the lab activities. One of the most important safety precautions that you can take is to wear personal protective equipment when it is required. While goggles can be uncomfortable, that discomfort is better than losing an eye. Always wear your goggles anytime you are working with glassware, chemicals, or projectiles. For lab activities that have safety precautions, you will find the appropriate safety icons immediately below the activity's equipment list. These icons are described below.

Body Protection—Chemicals or other materials could damage your skin or clothing. You should wear a laboratory apron, chemical-resistant gloves, or both.

Electricity—An electrical device (e.g., hot plate, lamp, microscope) will be used. Use the device with care.

Extreme Temperature—Extremely hot or cold temperatures may cause skin damage. Use proper tools to handle laboratory equipment.

Eye Protection—There is a possible danger to the eyes from chemicals or other materials. Wear safety goggles.

Fire Hazard—A heat source or open flame is used. Be careful to avoid skin burns and the ignition of combustible materials.

Sharp Object—Use care with the equipment in this lab activity to avoid cuts from sharp instruments or broken glassware.

Since some lab activities have specific steps with unique safety challenges, special safety notes are occasionally placed within the activity in ***bold italic font*** to emphasize their importance.

LAB EQUIPMENT AND TECHNIQUES

Science involves both equipment and techniques that you may not be familiar with. To help, we have included appendixes with helpful information. Appendix B explains many of the techniques that you will need for the lab activities in this lab manual. It includes some of the specialized equipment that you will use in the lab activities. Appendix C provides basic instructions for using the Phyphox® app, which utilizes the built-in sensors on a smartphone to make it a mobile physics lab. Appendix D explains the basics of using the Tracker video analysis software, which allows you to collect data from videos. Video technology provides a quick way to collect a lot of data and leaves more time for analysis. Appendix E is the same Mathematical Reference (Appendix B) from the Student Edition but is repeated here as a handy reference for use during lab activities.

INQUIRY LAB ACTIVITIES

The inquiry lab activities in this lab manual include tasks that are part of being a student physicist. They give you an opportunity to develop your own experiment to answer a scientific question. You will ask questions, form hypotheses, design investigations, collect and analyze data, draw conclusions, communicate results, and often develop additional questions—tasks that scientists do every day.

STEM ACTIVITIES

The STEM (Science, Technology, Engineering, and Mathematics) activities in this lab manual place you in the role of engineer as you are faced with a specific problem. You are tasked with designing something to solve a problem within stated constraints. In these activities you will follow the engineering design process: ask questions, imagine and plan solutions, build and test devices, evaluate solutions, and, if necessary, redesign, retest, and reevaluate a design. Appendix F outlines the engineering design process.

WRITING FORMAL LAB REPORTS

Communication is a key facet of science. Scientists need to communicate their findings to other scientists and ultimately to the world. Student scientists do this through writing formal lab reports, which you will write routinely if you take a college-level science course. Appendix G describes how to write a formal lab report. Use this to prepare your reports and to prepare for future science endeavors.

LET THE YEAR BEGIN!

One of the pleasures that we get from writing this book is knowing that for many of you the laboratory will become a special place. We invite you to come into the laboratory part of this course expecting to learn new and fascinating things. Look forward to the challenge of getting your hands on science. Your visits to the laboratory could be life-changing for you.

Welcome!

LAB
1A

Name ____________________

Date ____________________

Long Shot

Creating Histograms

The process of scientific inquiry hinges upon testing hypotheses and analyzing the results to see whether changes in the independent variable lead to statistically significant changes in the dependent variable. We often collect data and lots of it. Once we have data, we use various methods of analysis so that we can come to a conclusion. Does changing the independent variable affect the dependent variable? How much of a change is needed to imply a causal relationship? Scientists are interested in whether there is a statistically significant difference in different data sets. But what exactly is a statistically significant difference, and how can we quantify it? In this lab activity we will collect data and investigate the concept of statistical significance.

1. What is data? What types of data are there?

____

____

____

____

____

How can I tell whether two sets of data are substantially different or the same?

QUESTIONS

- How do I find the mean and standard deviation of a data set?
- How do I create a histogram?
- How can I tell whether the variable I introduced makes a difference?

EQUIPMENT

- computer with spreadsheet program
- metric tape measure
- computer paper (3 pcs.)

TABLE 1

Circumference (cm)
14.7
14.8
14.8
14.9
14.9
14.9
14.9
15.0
15.0
15.0
15.0
15.0
15.0
15.1
15.1
15.1
15.1
15.2
15.2
15.3

Data Table

It is impossible to always obtain the same measurement for the same object since measurements have random variations. This variation can be represented in a data table. The circumference of a sphere was measured twenty times using an accurate ruler, and the data is in Table 1. Note how the table is formatted.

- The top of the column has a label, including the units used.
- Since all the measurements were made with the same ruler, all the numbers are measured to the same decimal place.

2. What do you think is the benefit of including the units in the column header?

____

____

____

____

Histogram

Another way to represent data is with a graph called a *histogram*. To make a histogram, we first place the data into equal-sized bins and then count the number of measurements in each bin. The sizes of the bins are dictated by the data. We want to pick a value that will generate a definite mode, that is, a highest frequency, for the data. For example, the data in Table 1 can be binned every 0.1 cm. Table 2 includes the binned data. A histogram is a bar graph with the bins along the x-axis; the y-axis is the frequency of each bin value occurring. The associated histogram of that binned data is shown below. The shape of the graph is typical for experimental data and is often referred to informally as a *bell curve*. In the branch of mathematics called *statistics*, this shape is called a *normal distribution*. The graph has a peak near the middle and tapers off evenly on both sides. This shape makes sense because measurements have random variations but tend to cluster around a particular value.

TABLE 2

Circumference Bin (cm)	Number
[14.7–14.8)	1
[14.8–14.9)	2
[14.9–15.0)	4
[15.0–15.1)	6
[15.1–15.2)	4
[15.2–15.3)	2
[15.3–15.4)	1

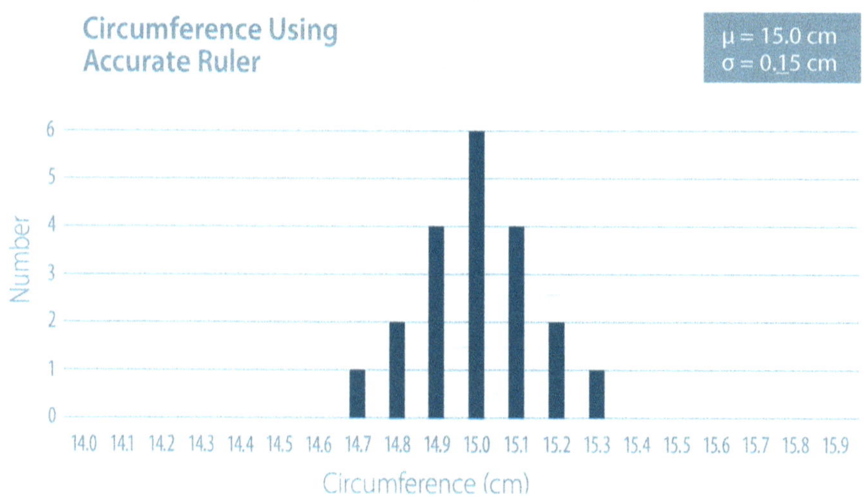

Name ______________________

Mean (or Average)

It is often convenient to have a single number that represents a data set. Physicists often use the *mean* of a set of measurements to represent the set. The mean (or average) is the sum of the data point values divided by the number of data points.

$$\mu = \frac{\Sigma x}{N}$$

In this formula μ is the mean, Σx is the sum of all the data points, and N is the number of data points. Spreadsheets make it easy to calculate a mean. The mean for the circumference of the sphere described above is 15.0 cm. Note that the mean is at the peak of the histogram. In most data sets, the mean is near the peak of the set's histogram.

Standard Deviation

When we use the mean to represent a data set, we lose a lot of information. Since the mean can represent widely spread data, it helps to have a measure of the distribution of the measurements around the mean. Are they mostly close or widely spread out? *Standard deviation* (σ) is a statistic that provides that information. Again, spreadsheet programs can easily calculate the standard deviation of a set of measurements since this is one of the many built-in functions of all spreadsheet programs. The standard deviation for the circumference data on page 2 is 0.15 cm.

Range

The range of a data set lists the minimum value through the maximum value. For the data in Table 1 the range is 14.7–15.3.

Reporting Measurements

Physicists will often represent measurements using both their mean and standard deviation. We could represent our measurements for the circumference as 15.0 cm ± 0.2 cm. The first number is the mean. The second number is the standard deviation. Recall that the standard deviation for our data is 0.15 cm, but we will use 0.2 cm as the standard deviation since our measurements were to the tenths place. The mean and standard deviation have the same number of decimal places.

Surveyors spend their days making measurements and then recording them on various maps; they must be both accurate and precise.

A measurement of 15.0 cm ± 0.2 cm means that most of the measurements were between 14.8 cm and 15.2 cm.

3. What do you think would happen if we measured the same sphere with a different ruler that was less accurate than the first ruler? What would happen to the histogram? What would happen to the mean and standard deviation?

____

____

____

____

____

Same Sphere, Different Ruler

Now let's assume that we've carried out the suggestion from Question 3. We've measured the same sphere twenty times using a less accurate ruler, yielding the histogram shown below.

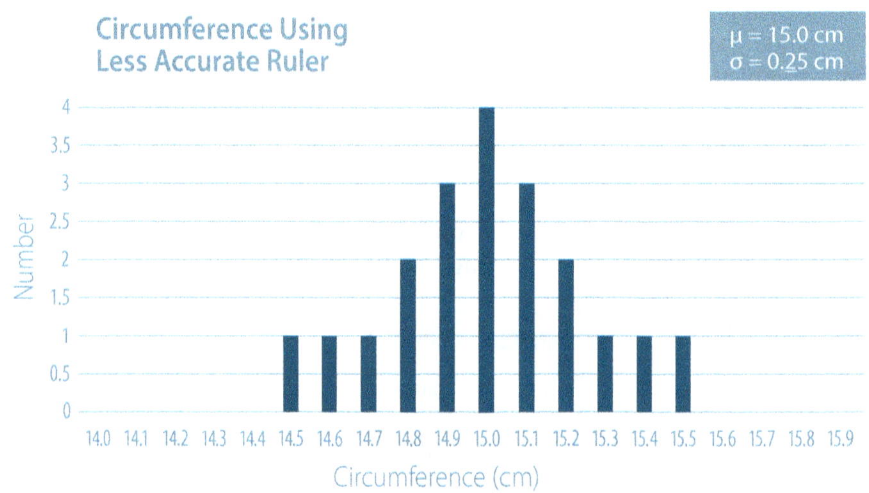

This set of measurements has the same mean but is more spread out. The peak is also not as high. We could represent this set of measurements for the circumference as 15.0 cm ± 0.3 cm. We can have more confidence in the measurements done with the more accurate ruler because of the smaller standard deviation of that set.

Hypothesis Testing

We can use the statistics discussed above to compare experimental groups to control groups after an experiment. We'll show how this works by using a series of graphs comparing different sets of data from experimental groups (orange) with data from a control group (blue) collected by different scientists working on the same experiment.

Name ______________________

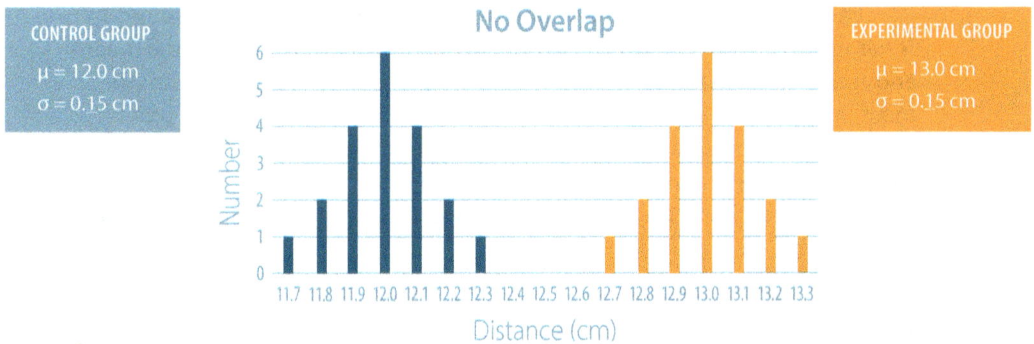

A quick visual inspection of the two histograms above shows that the experimental group *is different* from the control group. We can also show this by comparing the means and standard deviations. The control group is 12.0 cm ± 0.2 cm, which means that most of the data lies between 11.8 cm and 12.2 cm. The experimental group is 13.0 cm ± 0.2 cm, which means that most of the data lies between 12.8 cm and 13.2 cm. These two data sets do not overlap. The mean of the experimental group (13.0 cm) doesn't fall anywhere close to the standard deviation range of the control group. Therefore, we can make a strong argument that the experimental group is different from the control group.

The two histograms below have slight overlap, showing an experimental group that *is probably different* from the control group. We can also conclude

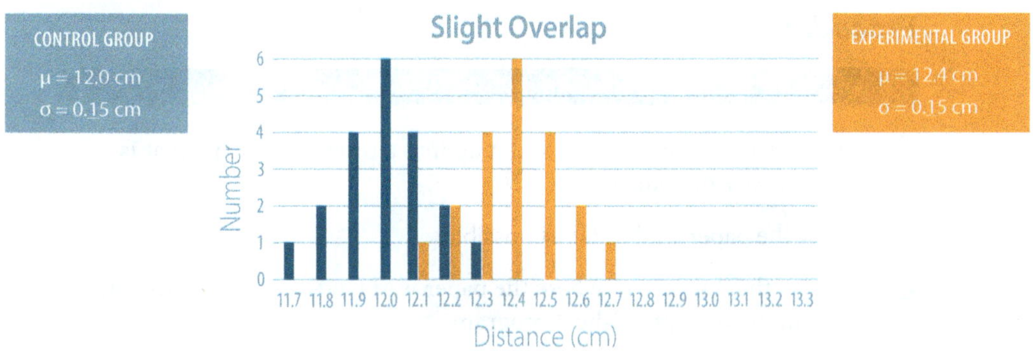

this by again comparing the means and standard deviations. The control group remains the same. The experimental group is 12.4 cm ± 0.2 cm, so most of the data lies between 12.2 cm and 12.6 cm. These two data sets overlap some, and the mean of the experimental group (12.4 cm) is close to, but not within, the standard deviation range of the control group. Therefore, we can make a case that the experimental group is probably different from the control group.

A visual inspection of the two histograms on the right shows that the experimental group *might be different* from the control group. Again, we'll look at the means and standard deviations. The experimental group is 12.2 cm ± 0.2 cm; most of the data lies between 12.0 cm and 12.4 cm. The data sets overlap more, and the mean of the experimental group is within the standard deviation range of the control group. So we could argue either way, that the experimental group either is or is not different from the control group. Any stronger conclusion would require the data to be analyzed using more sophisticated statistical techniques.

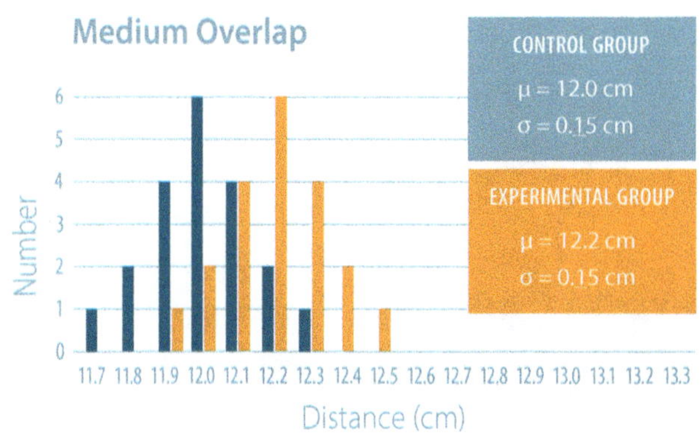

The last set of histograms (below) suggests that the experimental group *is not different* from the control group. We can verify this by using the means and standard deviations. The experimental group is 12.1 cm ± 0.2 cm, so most of the data lies between 11.9 cm and 12.3 cm. These two data sets overlap quite a bit. Also, the mean of the experimental group (12.1 cm) is well within the standard deviation range of the control group. We can make a strong case that there is no difference between the experimental and control groups.

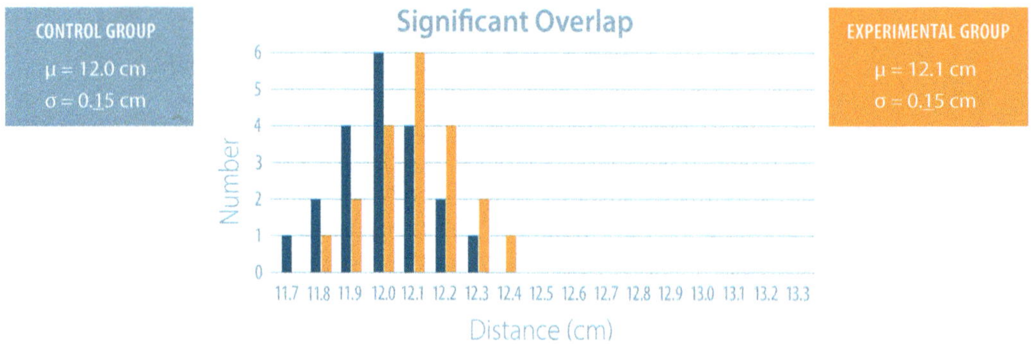

We will use this method of analysis to answer this question: Does a two-sheet paper wad go farther when thrown than a one-sheet paper wad?

Procedure

ONE-SHEET PAPER WAD

A Take one sheet of paper and wad it up into a rough ball shape that is about the size of your fist.

B Throw the paper wad as far as possible.

C Measure the distance to where the paper wad lands and record the measurement in your spreadsheet program.

D Repeat Steps B and C for at least 100 trials. Try to throw the paper wad the same way each time.

4. What is the distance thrown for the one-sheet paper wad? Express your answer as a mean ± standard deviation.

____

Name ____________________

TWO-SHEET PAPER WAD

Now you will repeat the experiment using a wad made from two sheets of paper.

5. How do you think making the wad from two sheets of paper instead of one will affect the distance? Explain.

____

____

____

____

____

E Take two sheets of paper and wad them up into a rough ball shape that is about the size of your fist. Make it about the same size as your one-sheet paper wad.

F Repeat Steps B–D for the two-sheet paper wad.

6. What is the distance thrown for the two-sheet paper wad?

____

Analysis

G Create a table of binned data for each set of data.

H Use the binned data to create a histograms for each data set.

I Determine the mean, standard deviation, and range for each set of distance data and record these values in Table 3.

7. How did the standard deviations of the two sets of data compare? What does this mean about the distances? Why do you think that happened?

____

____

____

____

____

J Find the range of values within one standard deviation (add and subtract the standard deviation value to the mean) for each set.

K Create a third histogram that includes both sets of data.

Conclusion

8. Describe how much the two histograms overlap. Would you claim that the data is the same for the two trials?

9. Does your data support your expectation from Question 5? Explain.

10. How might you change the procedures to get better results if you were to repeat the experiment?

Going Further

11. Why is it not sufficient to just compare the means of the distance thrown? What advantage is there in using a histogram and standard deviation?

12. A model is a simplified representation of an object or phenomenon in the real world. How are the mean, standard deviation, and histograms models of phenomena?

TABLE 3

One-Sheet Wad Distance (cm)		Two-Sheet Wad Distance (cm)	
Mean		Mean	
Standard Deviation		Standard Deviation	
Range		Range	

LAB
1B

Name

Date

Zeroing In

Inquiring into Experimental Design

Part of any experimental design is eliminating unwanted variables so that they don't cause random fluctuations or trends in our data. Having more consistent data decreases the standard deviation and makes our conclusions stronger. In this lab activity you will try to refine the experimental design of Lab 1A to reduce variability and collect more consistent data.

This activity is intended to introduce you to inquiry lab activities, in which you design an experiment to answer a scientific question. Lab 1A provides the basic procedures, but you will have to modify them here. The goal is to produce more consistent data than what you obtained with the basic procedures. In future inquiry lab activities you will design the entire experiment.

Can I modify an experiment so that it will produce more consistent data?

QUESTIONS

- » How do I design an experiment?
- » What do I mean by more consistent data?
- » How can I use my data to demonstrate that my experimental design provided more consistent data?

EQUIPMENT

- computer with spreadsheet program
- metric tape measure
- computer paper (3 pcs.)

Procedure

PLANNING/WRITING SCIENTIFIC QUESTIONS

A Brainstorm with your lab group ways that you can modify the paper wad activity from Lab 1A so that the data is more consistent. Consider the following questions: What can you change to launch the paper wad with the same speed? What can you change to launch a paper wad at a more consistent angle? How can you ensure that you always launch the paper wad from the same height? What can you do to more consistently measure the distance each time? How will you know whether your procedures have improved the results?

Hypotheses are possible explanations of a phenomenon that is testable. A good hypothesis explains the phenomenon and describes how the data and analysis may support or refute your hypothesis. A good hypothesis for this lab activity will include how the experimental data and the control data (from Lab 1A) must compare to support your claim.

B Write a hypothesis about how you will know whether your new data set is statistically different from the data from Lab 1A. Also include a statement about how you will know whether your new data set is more consistent than your data from Lab 1A.

DESIGNING SCIENTIFIC INVESTIGATIONS

C Write procedures to collect data that will allow you to test the hypothesis above.

D Have your teacher approve your procedures.

CONDUCTING SCIENTIFIC INVESTIGATIONS

E Follow the procedures that you have written and collect data to test your hypothesis.

DEVELOPING MODELS

F Use the statistical analysis process from Lab 1A, Steps G–K, to determine whether your one-sheet wad data is statistically the same as the one-sheet wad data from Lab 1A and whether the data is more consistent. Do the same analysis to determine whether your two-sheet wad data is statistically the same as the two-sheet wad data from Lab 1A and whether the data is more consistent.

SCIENTIFIC ARGUMENTATION

G State a claim, supported by evidence from your experiment, showing whether you were successful in improving the consistency of the data from Lab 1A. Whenever you state a claim, support the claim with evidence from your data or analysis.

LAB

2A

Name

Date

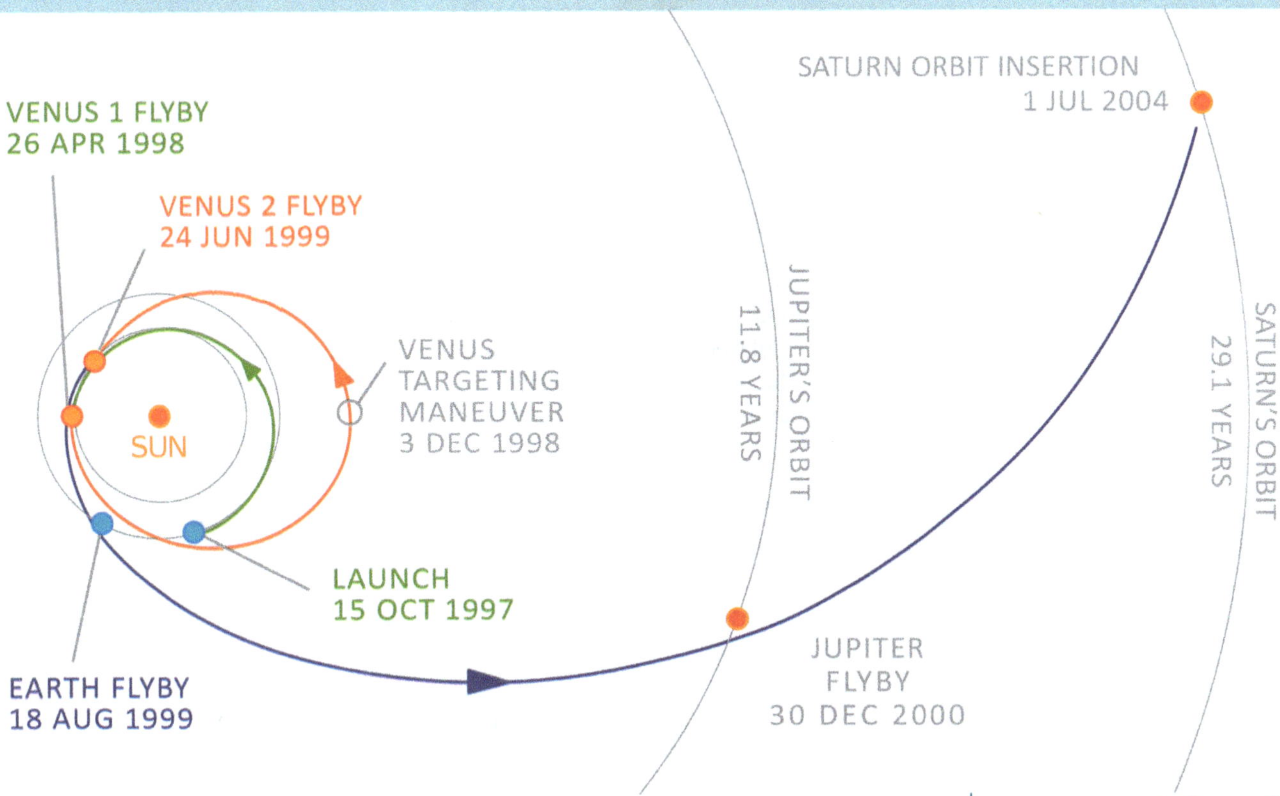

Keeping Things Rolling

Curve Fitting Using Video Analysis

Think about all the motion of the planets and moons in our solar system. Each planet orbits the sun at different speeds, and the speed of a particular planet even changes depending on where it is in its orbit. Then there are the moons which are all revolving around their planets. It is like a celestial ballet.

Recent NASA missions have involved a spacecraft flying close to numerous planets and moons. These near passes of the planets and moons allow for exploration of the bodies, but also assist the spacecraft's trips through, or even out of, the solar system. But how do physicists at NASA know when to launch a spacecraft and how to program it so that it passes close, but not too close, to these celestial objects? They do it with models, and more specifically, mathematical models.

What if we want to analyze the motion of something a little simpler, like a dragster in a drag race? We would begin by collecting data for the car. From the data we could produce mathematical models (graphs and equations) that represent the motion of the car. We would find the best model by using a technique called *curve fitting*. We could make predictions from the models and then test to see whether the models were accurate. In this lab activity you will analyze motion so that you can learn and practice this process of curve fitting.

How can I describe an object's motion?

QUESTIONS

- How can I collect sufficiently accurate position data for a moving object?
- What is the relationship between position, velocity, and acceleration?
- How can I analyze video data of a moving object?

EQUIPMENT

- video camera
- computer with video analysis and curve-fitting software
- ramp
- protractor
- meter stick
- rolling object (toy car or ball)

How to Do Curve Fitting

Imagine that someone rolls a toy truck down a ramp and takes measurements of its velocity. The data could be displayed on a graph called a *scatterplot*, shown as Graph 1. The purpose of a scatterplot is to determine the mathematical relationship between two variables.

Graph 1

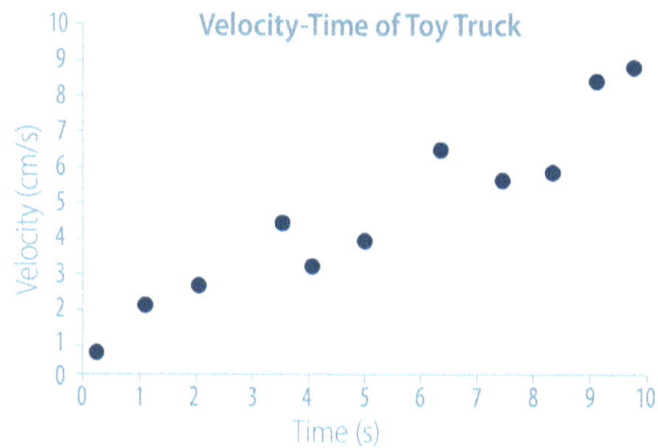

When we look at the general trend of the data points, we can see that they are roughly linear. A line has an equation of the form $y = mx + b$, where m is the slope of the line and b is the y-intercept, or vertical intercept. We could put a ruler on the graph to estimate a line that comes close to describing the data. An example of a possible line for this data is shown in Graph 2. This line has a slope of 0.6 and a y-intercept of 1.0. How well does the line seem to describe the data? Several of the data points are close to the line, but four data points are not. What direction do you think the line needs to move to fit the data better? It should be possible to find a line that better describes the data.

Graph 2

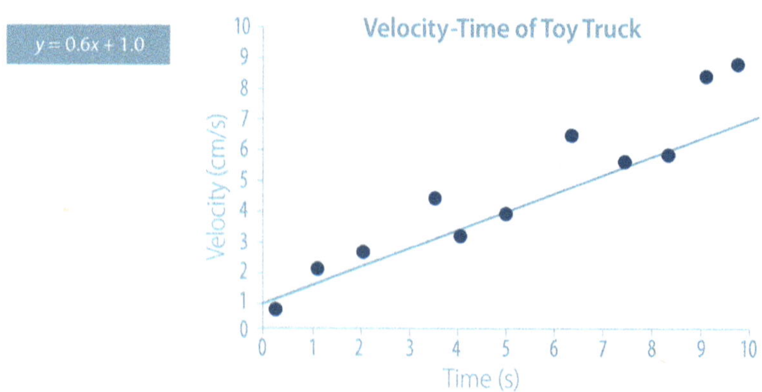

Name

One way to find a better-fit line is to find the sum of the vertical distances between the data points and the line. See Graph 3 for an example. We follow a procedure outlined below.

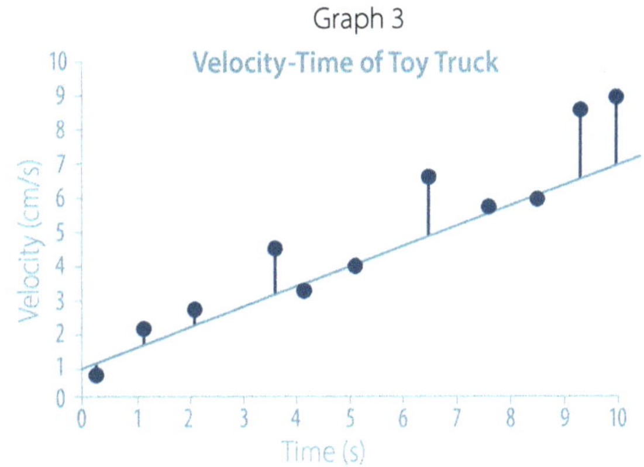

1. Change the slope of the line to one that seems to reduce the sum of the vertical distance.
   a. Find the new sum of the distances.
   b. Compare the new sum to the previous sum. If the new value is smaller than the previous value, keep the new slope.
2. Change the y-intercept of the line to again attempt to reduce the sum of the vertical distances.
   a. Find the new sum of the distances.
   b. Compare the new sum to the previous sum. If the new value is smaller than the previous value, keep the new y-intercept.
3. Repeat these steps until the sum is minimized.

As you can imagine, this process can be time consuming. Computers are good at repetitive calculations, so they are well suited for curve fitting. Graph 4 shows the best-fit line using Excel spreadsheet software. Notice that we have included $R^2 = 0.91$ below the equation for the line. R^2 is the *coefficient of determination*, a measure of the degree to which the independent variable determines the value of the dependent variable. If the independent variable perfectly determined the dependent variable, we would have a graph with $R^2 = 1$. In cases where the two variables are completely unrelated, the R^2 would have a value of zero. It is common for valid physics equations to have R^2 values between 0.9 and 1.0. Many curve-fitting programs may also report values for the *root mean square error* (RMSE) or *root mean square deviation* (RMS Dev). A good fit should have a very low RMSE. Similar curve-fitting processes can be used for other equations and data (including nonlinear relationships).

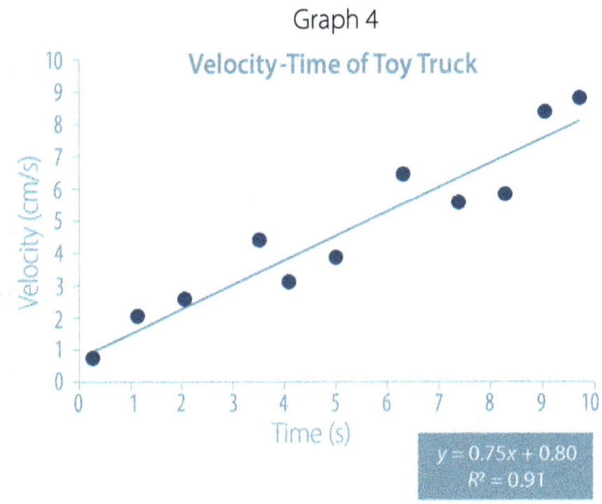

Now look at the equation of the line. The slope of the line is 0.75. For a velocity-time graph the slope is equal to the acceleration. Therefore, for this situation the acceleration is 0.75 cm/s². The y-intercept of the line is 0.80. This is the velocity when time is zero, or the initial velocity, which for this scenario is 0.80 cm/s.

In the example of curve fitting above, we were looking for the slope and y-intercept. These two values, which we adjusted to get the best-fit curve, are called *fitted parameters*. Physicists often extract useful information from fitted parameters.

Rule of Thumb

It is best to have at least five data points for each fitted parameter. With less than that our confidence in the best-fit equation is decreased. For example, if we are fitting an equation of the form $y = Ax + B$, there are two fitted parameters, A and B. Therefore, we would feel more confident if we had at least ten data points. As another example, if we are fitting an equation of the form $y = Ax^2 + Bx + C$, we would prefer to have at least fifteen data points. This is not an exact rule, but it is a good general practice. Sometimes we must fit a line using fewer data points because that is all the data available.

Video Data Collection Basics

A video of many scientific phenomena can be analyzed by tracking the position of moving objects within a frame of reference. A typical smartphone is capable of taking good videos for video analysis of motion.

When taking videos of physics experiments, it is important to consider the following.

- Record videos in a well-lit area. Brighter light will reduce motion blur.
- Orienting the camera parallel to the motion will reduce parallax error, which is due to the relative position of the observer, the phenomenon, and background information.
- Hold the camera steady when taking the video. Using a stand will be helpful.
- Place an object of known length at the same distance and the same plane as the motion within the frame. This will be needed to set the scale.

Name ______________________

Procedure

DATA COLLECTION

A Set up a ramp to produce a reasonable rate of motion for your rolling object.

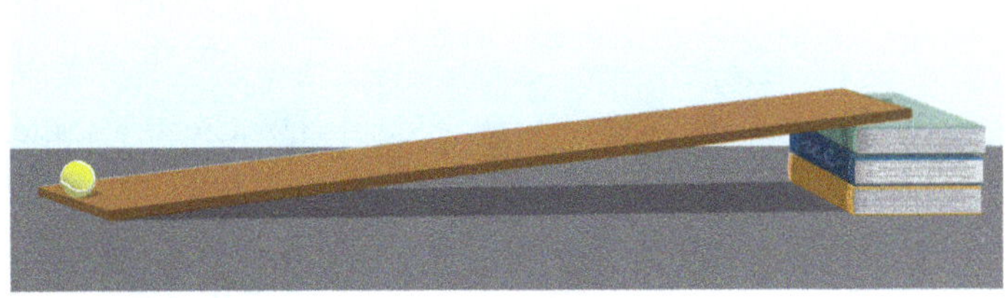

B Place the meter stick along the slope of the ramp so that it will not interfere with your rolling object. Make sure that it will be clearly visible in the video.

1. How do you expect the rolling object will move if you give it an initial velocity up the ramp?

____

____

2. Why do you think the rolling object will move as you expected in Question 1?

____

____

____

____

C Push and release your rolling object up along the ramp so that it goes a good way up the ramp before it comes back down.

D Position your camera so that you will be able to record the entire range of motion of the object. Record a video of the motion of the object.

E Repeat Step D to record the motion a few times to ensure that you will have at least one good video.

F Transfer the videos to a computer. It will be helpful if you rename the files to help you identify the experiment conducted.

VIDEO ANALYSIS

3. What do you expect will be the general shape of the graph produced by the data from the motion of your rolling object? Explain.

____

____

G Launch a video analysis program and analyze the videos. For basic help with Tracker video analysis software, see Appendix D.

H Set the origin at the base of the ramp.

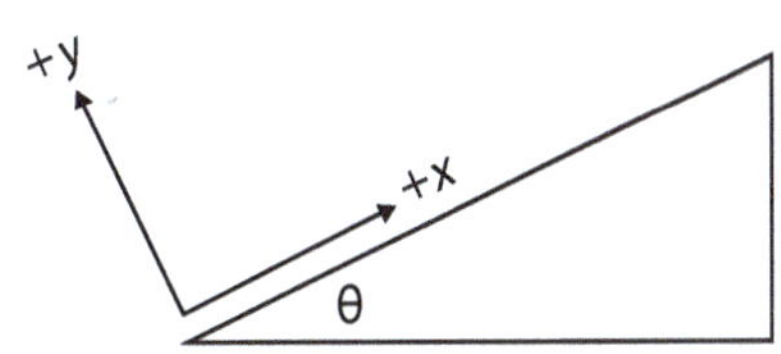

I Rotate the axes so that the positive x-axis is along the ramp.

J Set the scale using the meter stick that is in the video.

K Mark positions starting when the object is rolling on its own (no longer being pushed).

CURVE FITTING

Curve fitting can be done in the video analysis program, or the data can be moved to a spreadsheet program.

L Create a scatterplot of the x-position and time. Include a second-order polynomial (quadratic) best-fit curve.

M Create a scatterplot of the x-velocity and time. Include a first-order polynomial (linear) best-fit curve.

Analysis

4. Describe the shapes of the position-time and velocity-time graphs.

____

5. What is the equation for the second-order polynomial that is fit to the position-time graph? for the line that is fit to the velocity-time graph?

____

____

6. How well does the second-order polynomial fit the position-time data? How well does the line fit the velocity-time data? Use the R^2 value to explain your answers.

____

____

____

7. Calculate the acceleration from the equation from the second-order fit of the position-time graph. Explain how you determined this.

____

____

____

8. Calculate the acceleration from the linear fit of the position-time graph. Explain how you determined this.

____

____

____

The percent difference of two numbers is the difference between the numbers divided by the average of the two numbers, multiplied by 100%.

$$\%_{\text{difference}} = \left|\frac{x_2 - x_1}{(x_2 + x_1)/2}\right| 100\%$$

Name ______________________

9. Calculate a percent difference between the two values for acceleration. Show your work below.

Conclusion

10. Describe the motion of the object when its velocity is positive, zero, and negative.

____

____

____

11. Does the acceleration of the object change during its motion? How can you tell?

____

____

____

Going Further

12. How could the analysis in this lab activity be used to help people?

13. If you were able to minimize the friction between your object and the ramp, the only force that would cause the acceleration of your object would be gravity. Explain how you could use the acceleration of your rolling object and the ramp angle to determine the acceleration due to gravity.

LAB

2B

Name ______________________

Date ______________________

Time to Fall

Measuring Acceleration due to Gravity

For some people there is nothing more exciting than jumping out of an airplane! Others can't imagine taking such a leap. For a skydiver, the moment he leaves the aircraft he begins accelerating toward the earth. As he accelerates, a new force works to decrease his rate of acceleration. He will reach a maximum velocity—terminal velocity—once the air drag balances the force of gravity. His initial acceleration was the acceleration due to gravity.

Today we know that Earth's acceleration due to gravity is approximately 9.81 m/s^2 near its surface. In this lab activity you will measure the time that an object takes to fall a known distance. From this time and distance data, you will calculate the acceleration due to gravity.

1. What is the second law of motion? What do each of the variables in the equation represent?

____

____

____

In this lab activity you will drop an object initially at rest and you will time how long it takes to reach the ground. Because the object is small and the heights are relatively short, the object will experience a constant acceleration—the acceleration due to gravity.

How can I determine acceleration due to gravity?

QUESTIONS

- How does automating data collection affect the accuracy of an experiment?
- How can I evaluate the effect of different experimental methods?
- Which method is best for measuring the time it takes an object to fall?

EQUIPMENT

- smartphone physics app
- metric tape measure
- small mass, 25–50 g
- metal pan
- stopwatch
- ring stand
- iron ring
- balloons (5)

2. Write the known values for the variables in the second equation of motion for an object initially at rest that is dropped from a height (h) to the ground.

____

____

3. Plug these known values into the second equation of motion and solve for **g**.

In this activity you will use a traditional stopwatch along with an acoustic stopwatch that uses the microphone of a smartphone in conjunction with an app to measure the time it takes for the mass to fall. You will calculate the acceleration due to gravity using both methods and then use the known value of **g** to evaluate which method is more precise and which is more accurate.

4. Make a prediction as to which method you expect to be more precise and which to be more accurate.

____

____

____

____

Procedure

PART 1: MANUAL STOPWATCH METHOD

A Select a height that is at least 1.0 m above the floor. Measure and record this height in Table 1.

B Drop the mass from the height selected in Step A. When the mass is dropped, announce that it is dropped. (The metal pan should be positioned directly below the mass to help your lab partner know when the mass lands.) Your partner should face away from you to time the mass's fall with a stopwatch (a phone app will work just fine). Timing should begin as soon as you announce the drop of the mass and should stop when the mass strikes the pan. Record the time in Table 1.

C Repeat Step B for four additional trials.

PART 2: ACOUSTIC STOPWATCH METHOD

Smartphones have built-in sensors, and there are numerous apps that allow you to access these sensors. We will use the acoustic stopwatch in the Phyphox app for this portion of the lab activity. The acoustic stopwatch uses the microphone to start and stop the stopwatch.

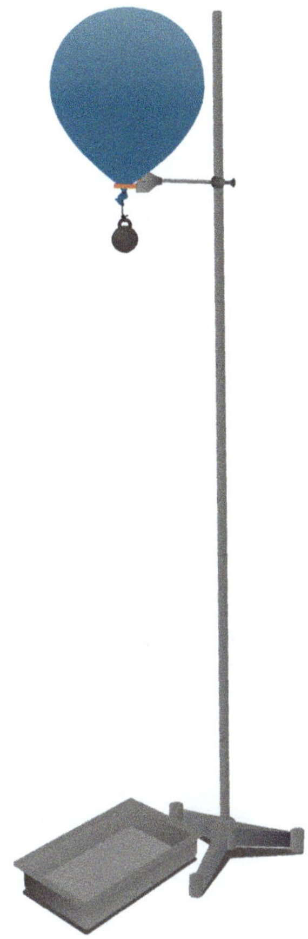

D Inflate the five balloons so that they are all approximately the same size.

E Attach the mass to an inflated balloon and set it up on a ring stand so that it looks something like the image at right. (You may want to use a binder clip to connect the mass to the balloon.) Position the metal beneath the mass. Adjust the setup so that the height of the mass above the pan is equal to the height used in Part 1.

F Launch Phyphox and select the acoustic stopwatch. This experiment will use the simple setting. To activate the acoustic stopwatch, press the start command (triangle) in the orange bar at the top of the app.

G Position the phone so that it is equidistant from the mass and the pan.

H Pop the balloon. The sound of the balloon popping will start the timer, and the sound of the mass striking the pan will stop the timer. Record the time in Table 1.

I Repeat the process for an additional four trials, recording each fall time in Table 1. You will need to verify the drop height for each trial due to variation in balloon size and mass attachment.

J Calculate the mean and standard deviation of the times for each timing method. The easiest way to do these calculations is with a spreadsheet program or the statistical features of a calculator. Record these values in Table 1.

K Use the mean time of each method to calculate the acceleration due to gravity. Record these values in Table 1. Show your work for the manual stopwatch below.

Analysis

5. Calculate the percent difference between the two values for **g** and record your calculation in Table 1. Show your work below. How do the two calculated values of **g** compare to each other?

6. Calculate the percent error for each method and record your calculation in Table 1. Show your work for the manual stopwatch below. Which calculation of **g** is closest to the accepted value of **g**?

Name ______________________

7. Which measurement method had the smallest standard deviation? Explain why you think this is true.

8. What factors contributed to the standard deviation of each method?

9. Why did you need to place the phone about halfway between the floor and the mass for the acoustic stopwatch method?

Conclusion

10. What additional steps to the procedure would improve the manual stopwatch method?

11. What additional steps to the procedure would improve the acoustic stopwatch method?

12. If you were to design an experiment, do you think that you would want to use manual techniques or automated techniques? Explain.

Going Further

13. This experiment assumed that objects accelerate once they are dropped. Suppose a friend claims that objects fall at a constant speed once they are dropped. How would you use the techniques from this lab activity to demonstrate that objects accelerate once they are dropped?

14. This activity shows how manual timing has a large degree of variability (higher standard deviation) than automated methods. How could you find a good representative value in cases where you had to use manual measurements?

TABLE 1

Height (m)		
Trial	**Manual Stopwatch Time (s)**	**Acoustic Stopwatch Time (s)**
1		
2		
3		
4		
5		
Mean Drop Time		
Standard Deviation of Drop Time		
	g	
Acceleration due to Gravity (m/s^2)		
Percent Difference		
Percent Error		

LAB
2C

Name

Date

Everyday Accelerations

Measuring Everyday Accelerations

Many people love the excitement of a roller coaster. And while most people don't get excited about the physics of a roller coaster, it is just that—physics—that produces the excitement in a roller coaster. Roller coasters have large forces, high velocities, and rapid accelerations. We experience forces, velocities, and accelerations every day, though not as drastically as on roller coasters.

In this lab activity you will use your knowledge of everyday motion to predict acceleration patterns in each dimension. Then you will use a smartphone app to measure the acceleration in each dimension. You will graph the results and evaluate whether the results matched your predictions.

Procedure

Smartphones have built-in sensors, and there are numerous apps that allow you to access these sensors. This lab activity will make use of the three-axis accelerometer that is part of a smartphone app. The procedures are written for the Phyphox app, but other physics apps will have a similar accelerometer utility.

How much acceleration do I experience in normal life?

QUESTIONS

» How can I measure acceleration during everyday motion?

» What do graphs of accelerated motion look like?

» How do scientists communicate their findings?

EQUIPMENT

- smartphone with physics app

PART 1: SETTING A COORDINATE SYSTEM

A Launch Phyphox on a smartphone and select the acceleration function (preferably without **g**). For more on the Phyphox app see Appendix C.

1. What do you think the acceleration graph would look like if you were to shake the phone?

B Start data collection by pushing the start button.

C Shake the phone, and then stop collecting data.

2. Sketch the graphs from shaking the phone.

D Start collecting data again and experiment with the phone to determine the directions of the x-axis, the y-axis, and the z-axis.

3. Explain how you determined the directions of the axes.

4. Sketch the smartphone and include arrows for the x-, y-, and z-direction of motion.

PART 2: INVESTIGATING EVERYDAY ACCELERATION

E Choose a type of motion experienced by people in their daily routine that you can safely record and that you believe will include acceleration.

5. Indicate the motion and predict what the graph will look like for each axis.

F Use your phone to measure this acceleration.

6. For two other common types of motion, indicate the motion and predict what the graph will look like for each axis.

G Repeat Step F for the other two types of motion.

Analysis

7. Describe how well you were able to predict the three forms of motion selected.

Conclusion

8. On the basis of the motion that you investigated in this lab activity, how much accelerated motion do you believe people experience in their daily lives? Explain.

Going Further

9. Research how much acceleration a human can take. What did you find?

10. Research the physiological responses of a human to excess upward (vertical) acceleration. What did you find?

11. How can our knowledge of how to measure acceleration be used in designing roller coasters, as well as aircraft and rockets, to be safe?

LAB

3A

Name

Date

Parts of the Whole

Investigating Vector Components

The roof on the house shown above has sections of varying slopes. A geometer might tell you that the upper roof sections are 1.46 m long and are sloped at a 24.1° angle. They would likely tell you that the lower left section of the roof is 2.46 m long with a 52.1° angle, while the lower right portion of the roof is sloped at 32.8° and is 5.53 m long. But this isn't how a builder would describe the roof. He would say that the upper portion of the roof is a 5.37:12 roof, while the left and right lower portions are 15.4:12 and 7.73:12 roofs, respectively.

While the geometer's measurements seem straightforward, the builders' measurements may seem puzzling. The builder's description of the roof sections are slopes listed as the rise over the run. The upper sections rise 5.37 m for every 12 m of run. If both descriptions are for the same roof sections, how are they related? In this lab activity we will investigate how vectors and their components are related both physically and mathematically.

How can I resolve a vector into components?

QUESTIONS

- What are vector components?
- How can component vectors define a vector?
- How does the field of trigonometry relate to real-world problems?

EQUIPMENT

- ring stand with clamp
- meter sticks (3)
- protractor
- level

Procedure

PART 1: DETERMINING COMPONENT VECTORS

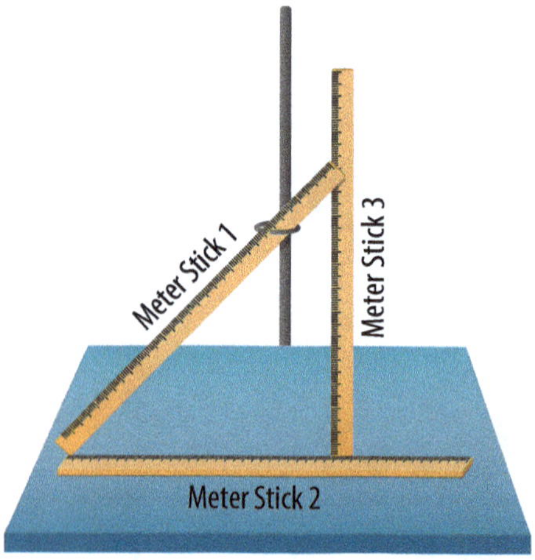

1. What are component vectors?

____

____

A Using the ring stand and clamp, set up one of the meter sticks as Meter Stick 1, as shown at left. One end of the meter stick should be touching the tabletop. Meter Stick 1 is your vector **M**.

B Measure the angle between Meter Stick 1 and the tabletop and record this value in Table 1.

C Lay Meter Stick 2 on the tabletop so that it aligns with the end of Meter Stick 1 that is touching the tabletop.

2. Which meter stick extends farther along the tabletop? Explain.

____

____

____

D Use the full length of Meter Stick 1 as your vector magnitude, $M = 100.0$ cm.

3. Estimate how much distance the vector covers along the tabletop.

____

E Stand Meter Stick 3 aligned with the raised end of Vector **M** and verify that it is vertical by using the level.

F Measure the x-component of the vector along Meter Stick 2 by measuring from the end of the vector to the location of Meter Stick 3 and record this value in Table 1.

4. How does your estimate in Question 3 match your x-component measurement?

____

G Measure the y-component of the vector along Meter Stick 3 by measuring from the tabletop to the elevated end of Vector **M**. Record this in Table 1.

H Reposition Meter Stick 3 so that it aligns with the 70.0 cm mark on Meter Stick 1. Then repeat Steps F and G for this new vector.

I Repeat Steps A through H, using a new angle for the vector, to obtain two more sets of data. You should now have data for Trials 1 and 2 at the original angle and for Trials 3 and 4 at the new angle.

Name ______________________

PART 2: FORMING VECTORS FROM COMPONENTS

5. How would two component vectors define a vector?

____

____

____

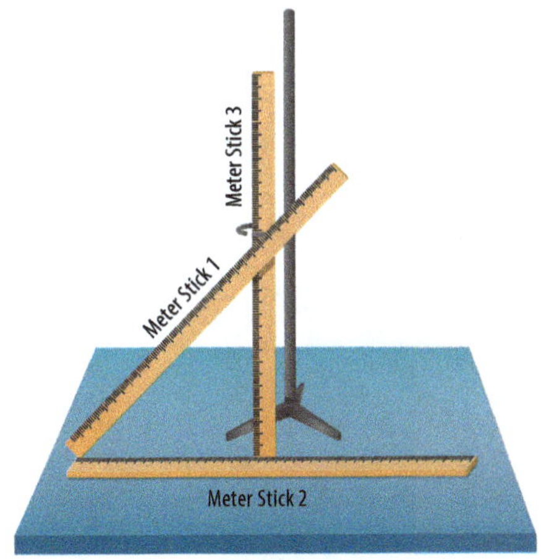

J Leave Meter Stick 2 on the tabletop and remove the other two meter sticks.

K Stand Meter Stick 3 at the 55.0 cm mark along Meter Stick 2. Use the ring stand and clamp and level to hold Meter Stick 3 vertically in place.

L Hold Meter Stick 1 (vector **M**) from the end of Meter Stick 2 so that it intersects Meter Stick 3 at the 47.0 cm mark. Measure the magnitude and angle of Vector **M** and record your data in Table 2.

M Repeat Step L with an intersection of Meter Stick 3 at 35.0 cm.

N Repeat Steps K–M with Meter Stick 3 located at 22.0 cm along Meter Stick 2.

Analysis

6. Sketch the vector triangle for Trial 1 of Part 1.

7. Using your sketch from Question 6, write the trigonometric equations that you would use to solve for the *x*- and *y*-components of the vectors from Part 1.

8. Calculate the x- and y-components of the vectors from Part 1. Show your work below for Trial 1, Part 1.

9. Sketch the vector triangle for Trial 1, Part 2.

10. Using your sketch from Question 9, write the trigonometric equations that you would use to solve for the magnitude and angle of the vectors in Part 2.

Name ______________________

11. Calculate the magnitude and direction of the vectors from Part 2. Show your work below for Trial 1, Part 2.

Conclusion

12. Why is trigonometry so important when working with vectors?

____

____

____

____

Going Further

13. Explain how the two descriptions of the roof relate to this lab activity.

____

____

____

14. How would a builder describe the roof shown below?

____

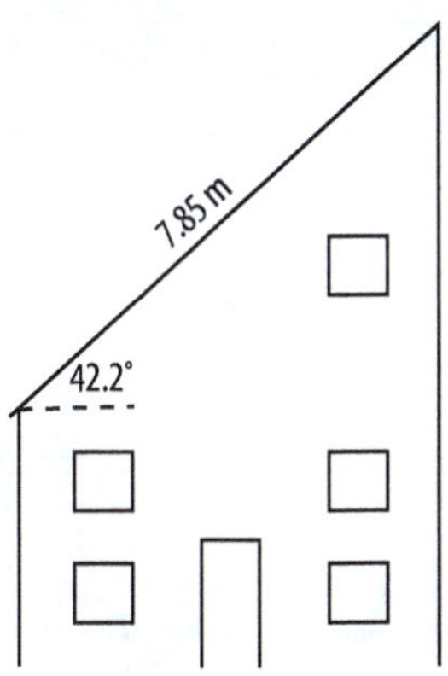

TABLE 1 *Component Vectors from Vectors*

Trial	Angle (°)	Magnitude (cm)	*x*-component (cm)	*y*-component (cm)	*x*-component (cm)	*y*-component (cm)
			Measured		Calculated	
1		100.0				
2		70.0				
3		100.0				
4		70.0				

TABLE 2 *Vectors from Component Vectors*

Trial	*x*-component (cm)	*y*-component (cm)	Magnitude (cm)	Angle (°)	Magnitude (cm)	Angle (°)
			Measured		Calculated	
1	55.0	47.0				
2	55.0	35.0				
3	22.0	47.0				
4	22.0	35.0				

LAB

3B

Name ____________________

Date ____________________

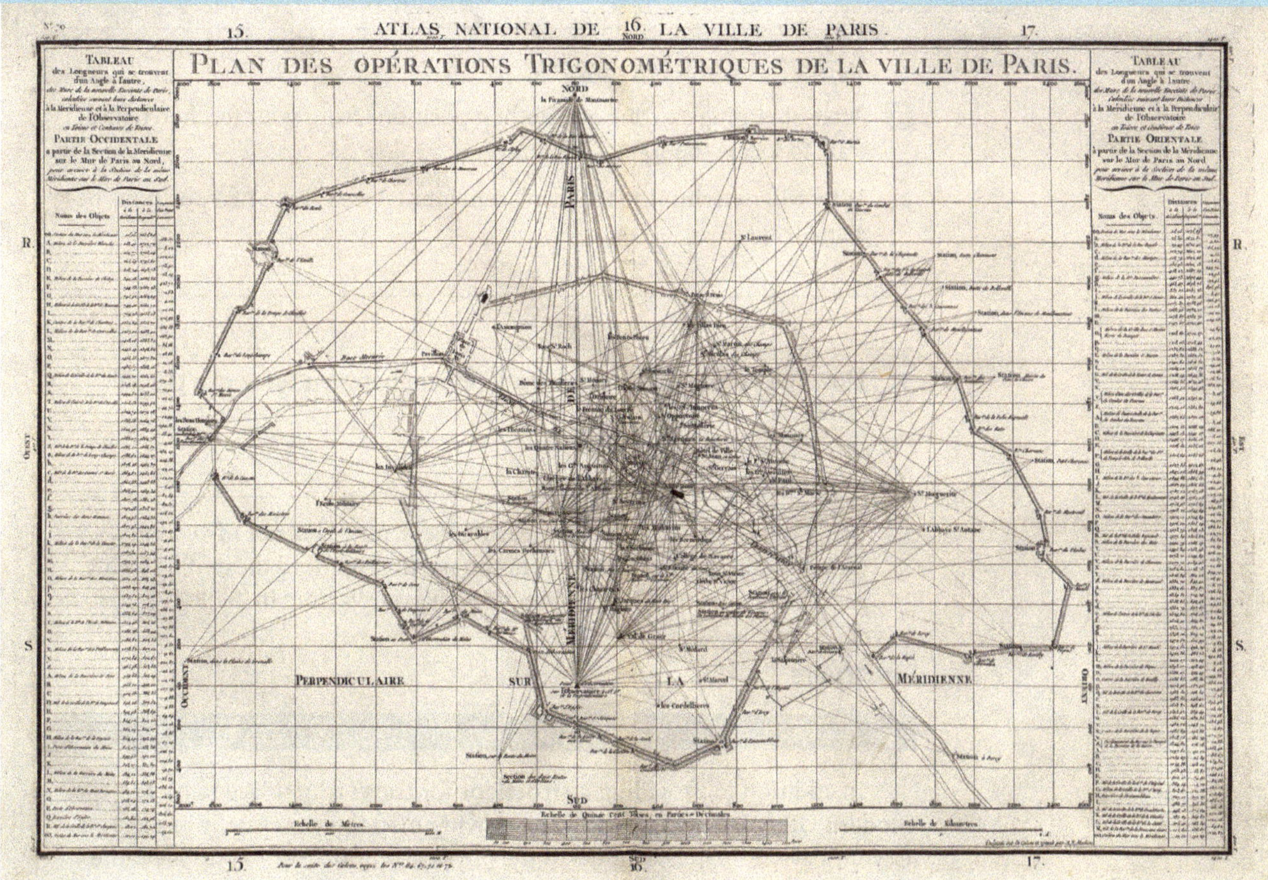

The Roundabout Way

Adding Vectors

Most of us can't imagine the days of discovery when people traveled the world without a map. Those early explorers often created maps as they went along that were a tremendous help for later travelers. Today most of us don't leave home without a smartphone map app. Through the wonders of the Global Positioning System (GPS) we have an ever-present resource to navigate the world. To make GPS available, scientists created a grid system on the earth using latitude and longitude to act as a frame of reference, like a coordinate plane, for navigation apps. Today airliners use GPS to navigate directly between any two points on the globe. NASA engineers use a similar coordinate system to direct robots to explore the surface of Mars.

In this lab activity you are going to use a series of vectors to represent a trip between two points around your school. You will then use graphical and algebraic methods to determine the displacement—that straight line path—between the starting point and the ending point.

How can I determine the displacement between two positions when I can't measure it directly?

QUESTIONS

- How can I create directions between two points?
- How can I use math to determine a measurement that I can't measure directly?

EQUIPMENT

- magnetic compass
- metric tape measure
- computer paper (1 pc.)

Procedure

1. Define *displacement.*

A After obtaining starting and ending positions from your teacher, take a walk to locate them.

2. Why will determining the displacement between these two positions be a challenge?

3. How can you determine the displacement between these two points?

B Using the compass and tape measure, determine a series of vectors that will allow you to travel from the starting point to the ending point. Record the magnitude and direction for your vectors in Table 1.

Analysis

GRAPHICAL APPROACH

In addition to providing instructions for travel from place to place, modern GPS systems also include a moving map to help visually orient you to your path. A map is a model of the physical world. You will do the same thing by drawing a scale map of your path.

C Determine the scale needed to complete your map on a single piece of computer paper.

4. What is the scale for your map?

D Create the scale model of your path.

5. What was the total distance traveled?

The scale model (map) that you made is the same as solving a vector addition problem using the graphical approach.

6. What is the displacement from the starting point to the destination on your map?

7. What is the displacement from the starting point to the destination in the real world?

Name ______________________

ALGEBRAIC APPROACH

The algebraic approach to vector addition problems uses trigonometry to determine the components of the individual vectors. These component vectors are summed to find the component vectors of the resultant. Trigonometry is again used to resolve the resultant component vectors into a magnitude and direction for the resultant.

8. Sketch the vector triangle for each leg of your path.

9. Using your sketch from Question 8, write the trigonometric equations that you would use to solve for the *x*- and *y*-components of each leg.

10. Calculate the *x*- and *y*-components of each leg and record your calculations in Table 1. Show your work for Leg A below.

E Sum the component vectors in each direction and record your calculations in Table 2.

11. Sketch the vector triangle for the resultant vector ($\Delta\mathbf{x}$).

12. Using your sketch from Question 11, calculate the magnitude and direction of the resultant displacement vector ($\Delta\mathbf{x}$).

Name ______________________

Conclusion

13. How do your two displacements compare? Explain the differences.

____

____

____

____

Going Further

14. This lab activity had you determine displacement by an indirect method because you couldn't measure the displacement directly. Give an example of something that you think would require scientists to make such indirect measurements.

____

____

____

____

TABLE 1

Vector	Magnitude (m)	Direction (°)	*x*-component (m)	*y*-component (m)
A				
B				
C				
D				
E				
F				
G				

TABLE 2

Vector	*x*-component (m)	*y*-component (m)	Magnitude (m)	Direction (°)
Resultant Δ**x**				

LAB

4A

Name ______________________

Date ______________________

It's a Tossup

Analyzing Projectile Motion

Shooting a free throw in basketball should be easy. You shoot from the same spot and no one is around you trying to block your shot. But it still takes a lot of coordination. The ball must be thrown at a particular angle and height and with a particular velocity to end in a relatively small area to make it through the hoop. That is why basketball players practice free throws all the time.

Now think about how complex basketball becomes when you factor in running, jumping, and the actions of opposing players. When a player is running, she must coordinate her velocity with the velocity of the ball. Is she running toward the basket, away, or even laterally? She may be jumping. Is she going to release the ball as she goes up, goes down, or is at the apex of her jump? Of course, while she is shooting, the opposing players are trying to block the ball! Even if they don't succeed in blocking a shot, they can still limit some of the angles at which she can shoot the ball.

In a basketball shot the basketball moves in projectile motion, which is the two-dimensional motion that combines motion in the horizontal direction and the vertical direction. This combination produces an arced trajectory. In simple projectile motion the only force acting on the projectile in flight is the force of gravity acting downward. Therefore, the horizontal

How do basketball players shoot accurately?

QUESTIONS

- How can I collect and analyze large amounts of data about complex motion?
- What do mathematical models tell me about real-world motion?
- How well do my models work?

EQUIPMENT

- video camera
- computer with video analysis software
- meter stick
- ball

direction motion has a constant velocity, and the vertical direction motion has a constant acceleration. The two motions combine to form a parabolic path. While we know that air drag is always present in real-world motion, it is negligible for many examples of simple projectile motion. Basketballs and other projectiles can be studied using video analysis. In this lab activity you will analyze the motion of a ball undergoing projectile motion.

Procedure

1. Describe the motion of a ball during a free throw.

___

A Place a meter stick parallel to the camera in a position that will be easy to see in a video. Choose a background that is well-lit with good contrast with the ball so that it will be easy to see it when doing video analysis. Record a video of one person tossing a ball until it lands on the floor. Save the video file with a descriptive name.

B Repeat video recording a few times to make sure that you have at least one good video.

C Send the video files to a computer.

Analysis

D Perform video analysis on the ball in the video. Be sure to mark the length of the meter stick as your scale. Set the first frame with the ball out of the thrower's hands at $t = 0.0$ s. Mark points after the ball has left his hands until just prior to the ball striking the floor.

E Construct the following graphs, using the indicated best-fit equation.

Graph A1: a position-time graph in the x-direction (linear fit)

Graph A2: a position-time graph in the y-direction (quadratic, second-degree-polynomial fit)

Graph A3: a velocity-time graph in the y-direction (linear fit)

Graph A4: a y-position versus x-position graph (quadratic fit)

2. Record the fitted equation and R^2 values for each of the graphs.

___

___

___

___

___

3. Are the best-fit curve equations that you determined for your graphs good models for the data? Explain.

4. According to physics, why does it make sense that the fits used are good models for the data? Explain.

5. Refer to the second equation of motion. What physical constant is the coefficient of t^2 term in the equation for Graph A2 related to?

6. Explain what each coefficient and constant in Graphs A1–A3 tell us about the ball.

7. According to your graphs, what is the measured acceleration due to gravity?

8. Since we are assuming that air drag is negligible, the vertical acceleration of the ball should be the acceleration due to gravity. Calculate the percent error of your measured acceleration due to gravity.

9. How do the coefficients in the equation for Graphs A2 and A3 relate to each other? Explain.

10. What does Graph A4 represent?

Conclusion

11. A classmate tells you that he believes that the ball's acceleration at the top of its path is 0 m/s^2. Is he correct? Provide evidence from your graphs.

12. Is projectile motion, neglecting air drag, a good model for the ball you threw? Would it be for all sports balls? When would this model begin to deviate from real-world motion? Explain.

Name

Going Further

13. Can basketball players neglect air drag when shooting the ball? Explain.

14. Why must aircrews delivering humanitarian aid consider air drag when executing an air drop?

15. How do you think the graphs would change if air drag were not negligible?

LAB
4B

Name ____________________

Date ____________________

Look Up in the Sky

Evaluating the Effect of Air Drag

If you have watched sporting events, you are familiar with the graceful arc of a basketball jump shot or a punted football. Your math teacher might note that the ball follows a parabolic path that we could model with a quadratic function. But have you noticed that in a baseball game a long fly ball seems to start along a parabolic path but appears to come down along a much steeper path than it went up? This same effect is seen in badminton.

The world record for the fastest badminton birdie is 493 kph (306 mph) by Malaysian Tan Boon Hoeng. But because of the physical characteristics of a birdie, it slows down rapidly. Therefore, the flight path of a birdie is not parabolic. In this lab activity you will compare the flight path of a badminton birdie to the path of the ball in Lab 4A. You will report your findings in a formal lab report (see Appendix G).

Procedure

The purpose of experiments is to test a hypothesis or to answer a scientific question. In this lab activity we want to answer two questions: How does the flight path of a badminton birdie compare to the path of the ball that you used in Lab 4A? Why is the path of a birdie different? The answers to these questions will guide the design of the experiment, the analysis of the data, and the interpretations and conclusions. All this information will be communicated to other scientists in the form of a lab report.

How does air drag alter the path of a projectile?

QUESTIONS

» How will the graphs of projectile motion differ with and without air drag?

» How do scientists communicate their findings?

EQUIPMENT

- video camera
- computer with video analysis software
- meter stick
- badminton birdie

A Place a meter stick parallel to the camera in a position that will be easy to see in the video. Choose a background that is well-lit with good contrast with the badminton birdie so that it will be easy to see the badminton birdie when doing video analysis. Record a video of one person tossing a badminton birdie until it lands on the floor. Save the video file with a descriptive name.

B Repeat video recording a few times to make sure that you have at least one good video.

C Send the video files to a computer.

Analysis

MOTION OF THE BIRDIE

D Perform video analysis on the birdie. Analyze the object after it has left the hand and before it strikes the floor.

E Construct the following graphs, using the indicated best-fit equation.

Graph B1: a position-time graph in the x-direction (linear fit)

Graph B2: a position-time graph in the y-direction (quadratic, second-degree-polynomial fit)

Graph B3: a velocity-time graph in the y-direction (linear fit)

1. Record the fitted equation and R^2 values for Graphs B1–B3.

____

____

____

____

2. Are the fits that you used for your graphs good models for the data for the birdie? Explain.

____

____

____

____

____

____

____

Notice in these graphs that the R^2 value is still pretty good, but also notice how the model (best-fit curve) deviates to one side of the data. A model that fits well will have a high R^2 value, but deviations of the data points for the model will be random. Since we are seeing systematic deviations, we should look for a better fit.

F Update your analysis on the birdie. Produce the following graphs.

Graph B4: Update the Graph B1 position-time graph in the x-direction with a quadratic (second-degree polynomial) fit.

Graph B5: Update the Graph B2 position-time graph in the y-direction with a cubic (third-degree polynomial) fit.

Graph B6: Update the Graph B3 velocity-time graph in the y-direction with a quadratic (second-degree polynomial) fit.

3. Record both fitted equations and their rms dev, R^2, or RMSE values for each of the best-fit equations for Graphs B4–B6.

____

____

____

4. Consider Graphs B4–B6. Are the best-fit equations good models for the data of the birdie? Explain your reasoning using both the graphs and the equations.

____

____

____

____

5. The quadratic fit (Graph B4) is a better fit than a linear equation (Graph B1). What does that tell you about the motion of the birdie in the x-direction?

____

____

6. Consider Graphs B2 and B5. How well do the second- and third-degree polynomials fit the data?

____

____

____

____

____

____

____

____

____

If you were to choose between these two fits, which would you choose? This is a challenging question. We must be careful when doing this type of analysis. We will almost always find that a higher-degree polynomial fits the data better, but we usually stick with the model that has the fewest adjustable parameters as long as the fit is good enough. We will follow this principle unless factors indicate that a more complex model is needed.

Between Graphs B2 and B5, Graph B2 (quadratic) would be a good choice because it is good enough. But let's consider a comparison of Graphs B3 and B6.

7. Which is a better fit of the data: Graph B3 (linear) or Graph B6 (quadratic)? Explain.

Is the equation with Graph B3 or Graph B6 better? The second-degree polynomial is a better fit. Therefore, an acceleration exists in the vertical direction beyond what is expected from the force of gravity alone. This is a strong evidence for the effects of air resistance. This also implies that the cubic fit (Graph B5) is actually better than the quadratic (Graph B2) for the position-time graph in the y-direction.

Let's look at the y-position versus x-position of each object.

G Create the following graphs.

Graph B7: a y-position versus x-position for the birdie (quadratic second-degree polynomial fit)

Graph B8: a y-position versus x-position for the birdie (cubic, third-degree polynomial fit)

Name ______________________

If the only acceleration acting on the birdie is *g*, then a parabola will be a good fit to the data. If the acceleration acting on the birdie includes an acceleration due to air drag, then a higher-degree polynomial will be required to describe the data.

8. Which equations fit best?

____

____

____

____

____

____

____

____

____

9. What do the general shapes of Graphs A4 and B8 tell you about the trajectory of a low-velocity ball and a birdie? Does this match your experience playing or watching sports?

____

____

____

____

____

____

H Write a lab report explaining your findings from the data for Labs 4A and 4B. As you write your report, be sure to keep in mind the questions mentioned above. You are trying to describe and explain the differences in motion for the projectiles in the two lab activities. Most high-school lab reports follow the format shown below.

Title

Introduction (or abstract): What is the purpose of the lab activity? What are the questions that the experiment is trying to answer?

Methods/Procedures: a detailed description of the procedures used so that another scientist can repeat the experiment

Data: raw data organized in a table; calculated data either in columns of the data table or in a separate table

Analysis: calculations, graphs, and evaluation of the data

Conclusion: Did the data support or refute the hypothesis? What are the answers to the questions that the experiment was trying to answer?

For more information on writing formal lab reports, see Appendix G.

LAB

5A

Name

Date

Balanced Approach

Investigating Balanced Forces

Civil engineers plan projects for the construction of huge structures. As you can imagine, there are many forces on these structures. A significant part of the design process is to analyze the forces that will act on a structure. The bridge in the image above is made of numerous metal girders that allow the bridge to support itself and all the vehicle traffic that will travel across it. Each structural member will have either a tension force or a compressive force on it due to the interaction of all the other parts of the structure. As vehicles move across the bridge, the forces on each member change. In this lab activity we will use a force table to analyze systems consisting of a few forces.

1. What is a resultant?

____

____

How can I balance a system of forces?

QUESTIONS

» How can we measure forces on a system in equilibrium?

» How can I use information about forces in equilibrium to determine an unknown force?

EQUIPMENT

- force table and accessories (including 3 pulley clamps)
- laboratory balance
- metric mass set
- heavy string
- large bolt

2. A force of 30.0 N acts along the positive x-axis. Another force of 45.0 N acts 52.5° above the positive x-axis. Determine the resultant algebraically. Show your work in the space below.

Name ______________________

3. What force would balance the two forces in Question 2? In other words, what third force would put the system in equilibrium?

____

____

Procedure

SETUP

A Fasten three strings to the force table ring by tying a small loop in each piece of string.

B Place the ring over the force table's center peg.

C Thread the loose ends of the strings over the pulleys. Tie a loop at the end of each string.

TRIAL 1

D Arrange the pulleys so that String 1 is positioned directly over the 0° mark and String 2 is positioned over the 90.0° mark (counterclockwise from the 0° mark). Position the third pulley at approximately 180°.

E Attach 150.0 g to the free end of String 1 and 200.0 g to the free end of String 2. (See right.) Attach 250.0 g to String 3.

F Adjust the angular position of the third pulley to center the ring around the peg. Measure the direction of Force 3 and record your data in Table 1.

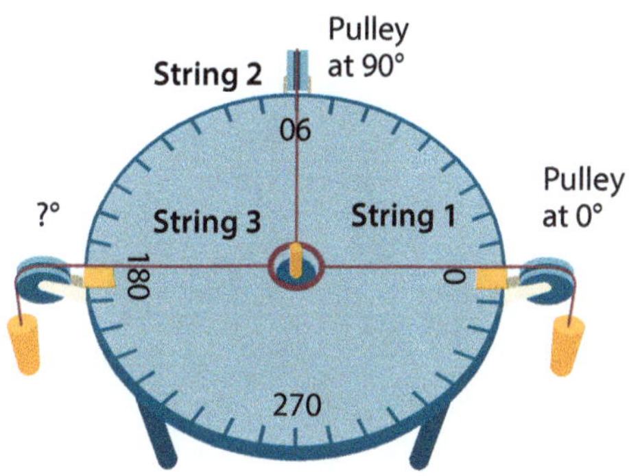

TRIAL 2

G Change the mass on String 1 to 200.0 g, keeping the pulley at 0°. Keep the mass on String 2 at 200.0 g and move the pulley to 55.0° (counterclockwise from 0°). Change the mass on String 3 to 350.0 g.

H Adjust the angular position of the pulley for String 3 left or right to center the ring over the peg. Record the direction of Force 3 in Table 1.

TRIAL 3

I Change the mass on String 1 to 150.0 g, keeping the pulley at 0°. Change the mass on String 2 to 250.0 g and move the pulley to 65.0° (counter-clockwise from 0°). Change the mass on String 3 to 350.0 g.

J Adjust the angular position of the pulley for String 3 left or right to center the ring over the peg. Record the direction of Force 3 in Table 1.

TRIAL 4: FINDING AN UNKNOWN

K Replace the mass on String 3 with the bolt and position String 3's pulley at 180°.

L Replace the masses on Strings 1 and 2 with appropriate mass (150.0 g–400.0 g should work). *Hint*: the sum of the masses on Strings 1 and 2 must be at least as much as the mass of the bolt. Record the masses for Strings 1 and 2 in Table 1.

M Adjust the angular position of the pulleys of Strings 1 and 2 to center the ring over the peg. Record the direction of Forces 1 and 2 in Table 1.

N Measure the accepted mass of the bolt with the laboratory balance and record your data in Table 3.

Analysis

O Calculate the magnitude of Forces 1, 2, and 3. Show your work from Trial 1 Force 1 in the space below and record your results in Table 1.

Name ______________________

4. Use the algebraic vector addition method to find the resultants of adding Forces 1 and 2 for Trials 1–3. Determine the opposite angle for the balancing force. Record your answers in Table 2. Show your work for Trial 2 below.

5. Do your measured and experimental values for Force 3 indicate that you were accurate in the lab activity? Explain.

6. Use the algebraic vector addition method to find the resultants (in vector component form: <$\mathbf{F}_{3x}$, $\mathbf{F}_{3y}$>)of adding Forces 1 and 2 for Trial 4.

7. Using your result from Question 6, determine the magnitude of Force 3 in Trial 4. Record this value in Table 3.

8. Using the magnitude from your answer to Question 7, calculate the measured mass of the bolt from the force table readings and record your result in Table 3.

Conclusion

9. Calculate the percent error between the measured mass (expected value) and the mass that you calculated for Trial 4. Record your percent error in Table 3.

10. Make a scientific argument (claim and evidence) for the effectiveness of this method for determining the mass of the bolt.

Going Further

11. Look back at the introduction to this lab activity and the accompanying photo. How do you think civil engineers analyze structures such as bridges? Explain.

Table 1

Trial	Force 1			Force 2			Force 3		
	Mass (kg)	Magnitude (N)	Direction (°)	Mass (kg)	Magnitude (N)	Direction (°)	Mass (kg)	Magnitude (N)	Direction (°)
1	0.1500		0.0	0.2000		90.0	0.2500		
2	0.2000		0.0	0.2000		55.0	0.3500		
3	0.1500		0.0	0.2500		65.0	0.3500		
4									180.0

Table 2

Trial	Magnitude (N)	Direction of F_3 (°)
1		
2		
3		

Table 3

Accepted Mass (g)	Calculated Magnitude (N)	Measured Mass (g)	$\%_{error}$

LAB
5B

Name

Date

Don't Be So Pushy!

Inquiring into Newton's Second Law

Today we recognize Newton's laws of motion as fundamental to the study of physics, but they were revolutionary in his time. Upon his foundation future scientists developed the modern science that ushered in the Industrial Revolution. Newton's second law of motion relates the concepts of force, mass, and acceleration. Newton's laws were a complete shift from Aristotle's concept of motion in which force was thought to be related to the velocity of the object. In this inquiry lab activity you will design an experiment to confirm Newton's second law.

What does motion caused by a constant applied force look like?

QUESTIONS

- How do we establish a constant force on a moving object?
- What do the motion graphs look like when an object moves under a constant force?
- How are mass, force, and acceleration related?

Procedure

PLANNING/WRITING SCIENTIFIC QUESTIONS

In the lab activity that you design you will investigate the concept of motion under a constant applied force. You will build an experiment in which you apply a constant force and then measure the motion (e.g., position, displacement, velocity, acceleration). You will specifically want to know how the mass, applied force, and acceleration of the object relate to each other when you apply a constant force to an object.

A Brainstorm with your lab group what you expect the motion of an object acted upon by a constant force to look like. Sketch an acceleration-time graph, a velocity-time graph, and a position-time graph. Remember to think about how the mass, applied force, and acceleration of the object relate to each other. Also think about how each of the graphs relate to each other.

B Brainstorm different approaches that you could use to set up this experiment. Think about the challenges to designing an experiment with a constant applied force. Also, make sure that you consider how you will collect your data.

C Write specific questions related to the motion of an object under a constant applied force that you will answer with your experiment.

DESIGNING SCIENTIFIC INVESTIGATIONS

D Write procedures to set up your experiment and collect data that will allow you to answer the questions that you wrote in Step C. Remember that to understand the relationship between two variables, you will need to change one variable (the independent variable) to measure the effect that change has on the second variable (the dependent variable) without changing any other variables.

E Have your teacher approve your procedures.

CONDUCTING SCIENTIFIC INVESTIGATIONS

F Following the procedures that you have written, collect the data to answer the questions that you wrote.

DEVELOPING MODELS

G Analyze your data to confirm that you were able to produce motion caused by a constant applied force.

H Analyze your data to evaluate the relationship between mass, force, and acceleration.

SCIENTIFIC ARGUMENTATION

I Answer any remaining questions that you wrote in Step C.

J State a claim about the relationship between the mass of the object, the applied force, and the acceleration of the object. Always support your claim with evidence from your experiment.

LAB
5C

Name ______________________

Date ______________________

Bridging the Gap

Designing a Bridge

Throughout history cities and towns have developed around navigable waterways. Countries tended to develop along the coasts first, then along the navigable rivers. As helpful as rivers are, they become a hindrance when we must cross them. Initially, humans spanned narrow streams with a fallen tree or found shallow regions across which they could ford. But large rivers required people to build bridges. Today we have bridges that span huge gaps. The longest bridge in the world is the Danyang-Kunshan Grand Bridge in China, measuring 164.8 km. The longest span of a bridge is the 1.99 km center span of the Akashi Kaikyō Bridge in Japan. In this STEM (Science, Technology, Engineering, and Mathematics) activity you will design a bridge. The goal is to build a bridge that will have the highest load-to-weight ratio.

How can I design and build an efficient bridge?

QUESTIONS

- What are the key elements to building a bridge?
- How can I maximize the efficiency (the load-to-weight ratio) of a bridge?
- How can I improve on my initial design?

Procedure

DESIGN PARAMETERS

As you design your bridge, you must work within the design parameters listed below.

Your teacher will specify the materials that can be used for your bridge.

Your bridge must be able to span an opening that is ________ cm, and the bridge must have a width between ________ cm and ________ cm.

Your bridge must be designed to facilitate the testing device, which your teacher will show you.

Your goal is to maximize the load-to-weight ratio of your bridge. In other words, we want the bridge to be efficient.

PLANNING THE DESIGN

A Research bridge designs. There are many computer programs and smartphone apps that can be used for bridge design. Do an internet search using the keywords "bridge design program," "bridge designer," or "West Point bridge designer."

B Keeping in mind the design parameters listed above, brainstorm possible designs for your bridge with your group.

C Draw and label a diagram of your plan for your bridge. Indicate on the diagram that this is a *preliminary design*.

D Share your research findings with your group. Reach consensus on a final design that your group will build and submit for testing.

E Have someone in the group draw a full-size side view of your bridge. You will use this during construction. Indicate on this diagram that it is an *initial production design*.

BUILDING THE BRIDGE

F Using the drawing that you created in Step E, build the two side structures of your bridge. A great technique is to cover the drawing with wax paper and use it as a template. The wax paper will prevent any glue from ruining the drawing.

G Connect the two side structures with horizontal connecting members. Be sure that the placement of the horizontal connecting members will not prevent the installation of the testing apparatus.

H Take pictures of your completed bridge. You will use these when you evaluate the performance of your bridge.

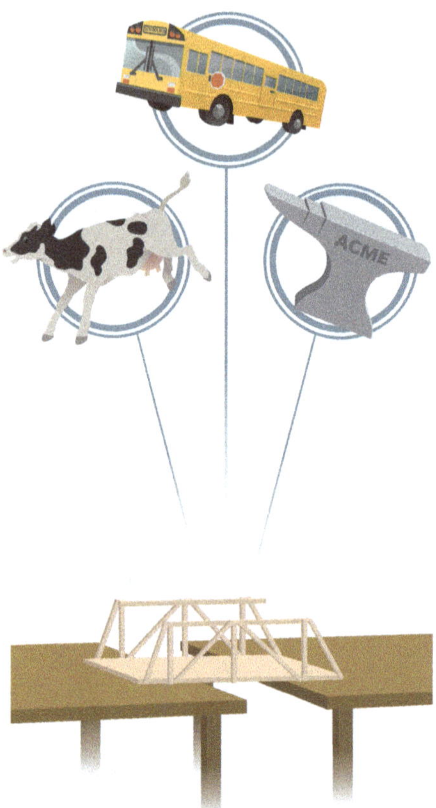

TESTING THE DESIGN

I Determine the weight of your completed bridge.

J Follow the procedure provided by your teacher for testing your bridge. Record the maximum load applied to your bridge prior to failure.

REFINING THE DESIGN

K Evaluate the performance of your bridge. Consider where and how your bridge failed (failure mode). Compare the failed bridge with the photographs that you took in Step H to assist in your analysis.

L Outline specific modifications that you could make to the bridge to improve its performance.

M If time permits, build your redesigned bridge.

RETESTING THE DESIGN

N If time permits, test your redesigned bridge.

LAB

6A

Name ______________________

Date ______________________

Slipping Away

Measuring Coefficients of Friction

One of the primary forces involved in operating a motor vehicle is friction. We often think that we want to minimize friction to improve the fuel economy of our cars. This is true when we are talking about the internal friction of the engine, the rolling friction of the tires, and the air drag on the car. But we also want to increase the friction between the tires and the road to increase traction. We want more friction when we need to stop our car in an emergency. And everyone wants enough friction to keep his car from sliding off the road when going around a turn. As we can see, friction is both good and bad when it comes to our cars. In this lab activity we will investigate the force of friction, both static and kinetic.

1. Define *friction*.

____

____

____

How does friction affect a car's operation?

QUESTIONS

» How are friction forces related to the normal force?

» How do we determine the coefficients of static and kinetic friction?

EQUIPMENT

- spring scales, 10 N, 20 N
- object to slide
- wood board
- protractor
- string

2. Complete the table below.

Term	Symbol	Definition
Static Friction		
Kinetic Friction		
Normal Force		
Coefficient of Static Friction		
Coefficient of Kinetic Friction		

Procedure

PART 1

A Use the spring scale to measure the weight of the object and record it in Table 1.

B Place the board on the tabletop and place your object on top of the board at one end of the board.

C Slowly raise the end of the board that is under the object until the object starts sliding down the board.

D While holding the board in place, have another member of the lab group measure the angle of the incline. Record this value in Table 1.

E Repeat Steps B–D three more times. Record the values for each trial in Table 1.

These skaters have to work against friction to keep moving, but friction also allows them to follow the curves of the road.

Name ______________________

PART 2

F Record the weight measured in Step A in Part 2 of Table 1.

G Lay the board on the tabletop and place your object on top of the board.

H Use the string to attach the spring scale to the object.

I Pull the object horizontally across the board at a constant speed. Do your best to keep the spring scale horizontal when pulling. Read the applied force from the spring scale and record it in Table 1.

J Repeat Step I three more times. For each new trial, pull in the opposite direction along the board.

3. Why do you think that you must pull the object in opposite directions?

____

____

____

Analysis

PART 1

4. Draw the free-body diagram for the object as it is on the inclined plane.

5. Using the free-body diagram from Question 4, set up Newton's second law for both the x- and y-directions. Calculate the magnitudes of the normal force and friction force for each trial and record them in Table 1. Show your work for Trial 1 below.

6. Use the force magnitudes calculated in Question 5 to determine the coefficient of friction for each trial and record the values in Table 1. Show your work for Trial 1 below.

Name

7. The coefficient of friction is related to the angle of the inclined plane. Use your work from Questions 5 and 6 to show which trigonometric function is related to the coefficient of friction.

PART 2

8. Draw the free-body diagram for the object as it is pulled along the board.

9. Using the free-body diagram from Question 8, set up Newton's second law for both the *x*- and *y*-directions. Calculate the magnitudes of the normal force and friction force for each trial and record them in Table 2. Show your work for Trial 1 below.

10. Use the force magnitudes calculated in Question 9 to determine the coefficient of friction for each trial and record the values in Table 2. Show your work for Trial 1 below.

Name ________________

11. How do the coefficients of friction from Parts 1 and 2 compare? Is any difference due to experimental error or to some characteristic of friction? Explain.

Conclusion

12. In Part 1 you lifted the board until the object started to move. In Part 2 you pulled the object across the board. If there was motion in both Parts 1 and 2, which part of the lab activity is designed to determine the static coefficient of friction? Explain.

Going Further

13. Give an example not mentioned in the lab activity of a situation in which having more friction is harmful and one in which having more friction is beneficial.

14. How does what you learned in this lab activity relative to static and kinetic friction apply to stopping a car in an emergency? Explain.

TABLE 1

Part 1

Trial	Object Weight (N)	Angle (°)	Normal Force (N)	Friction Force (N)	Coefficient of Friction
1					
2					
3					
4					

Part 2

Trial	Object Weight (N)	Applied Force (N)	Normal Force (N)	Friction Force (N)	Coefficient of Friction
1					
2					
3					
4					

LAB
6B

Name

Date

Putting It All Together

Investigating Combined Forces

For a pilot there is nothing more exciting than reaching the end of the runway and lifting off into the wild blue yonder. But there is a lot of analysis that occurs before that point. First, the pilot must calculate takeoff data to make sure that the aircraft will be moving fast enough by the end of the runway. The movement of aircraft is affected by its own weight and the weight of the crew, passengers, fuel, and cargo. The thrust of the engines pushes the plane down the runway, but thrust depends on air pressure and temperature. To get airborne the aircraft must produce enough lift. Lift is also affected by air temperature and pressure, as well as by the shape of the wing, which is modifiable by the pilot. Of course, there is friction and air drag working against the forward motion of the aircraft. In addition to all these factors, the pilot also must take into account whether the runway is sloped, whether there is a headwind or tailwind, and whether there is any rain or snow. While all these variables seem daunting, a pilot is trained to account for all these factors, and the reward is great once he is airborne.

In this lab activity we will analyze a system with fewer forces, but the basics of the analysis will be the same. We will analyze the motion of a dynamics cart as it is pulled up a ramp. The force propelling the cart is tension in a string. This tension is produced by a mass hanging over the edge of the ramp. The force of gravity pulling the mass downward starts the whole process.

How can we analyze the motion of an object with many forces on it?

QUESTIONS

- How can we collect and analyze data with videos?
- What causes variations in experimental data from expected data?

EQUIPMENT

- laboratory balance
- video camera
- computer with video analysis and spreadsheet software
- dynamics cart
- inclined plane
- support rod
- heavy-duty support stand
- C-clamp
- pulley clamp
- protractor
- meter stick
- string, 2 m
- masses, 500.00 g (3)

1. A 1.500 kg mass is suspended from a string. The string runs over a pulley and is attached to a 546 g dynamics cart that is resting on an inclined plane at an angle of 32.0° from the horizontal. The cart is released. Draw the free-body diagram of the cart at the instant that it is released.

2. Calculate the magnitude of the acceleration of the cart described in Question 1. Assume no friction.

Name ______________________

Procedure

SETUP

A Measure the mass of the cart and record it in Table 1 for Trials 1 and 2.

B Set up the inclined plane near the edge of the table as pictured on the right. Set the angle of the plane at 20.0° above the horizontal. The method for supporting the plane may be different than illustrated.

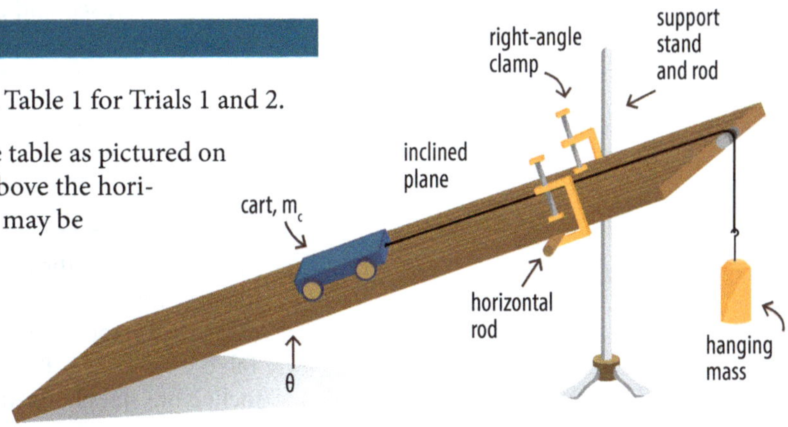

C Plan your video so that you can record the motion of the cart up the ramp. Place the meter stick in the video frame for setting the scale during video analysis.

D Choose a hanging mass for Trial 1 that is approximately equal to the mass of the cart. For most carts 500.00 g will work well. Measure the hanging mass and record this value in Table 1 for Trial 1.

E Tie a string to the mass. The string should be long enough to allow the mass to fall freely until the cart reaches the top of the inclined plane.

F Place something under the mass to keep it from marring the floor.

DATA COLLECTION

G Hold the cart steady in this position until it is ready to be released.

H Hook the string that is attached to the hanging mass to the front of the cart. Suspend the mass over the pulley and stop any swinging motion.

I When ready to run the trial, start recording the video.

J Release the cart. Do not impede or accelerate the cart when releasing it. ***If not stopped, the cart will hit the pulley and can fly off the end of the inclined plane. Someone in the lab group must be ready to stop it.***

3. What would happen to the magnitude of the acceleration if the hanging mass increased and everything else remained the same? Explain.

____

____

____

____

K For Trial 2, repeat Steps G–J using a hanging mass that is about 500.00 g heavier than the mass used in Trial 1. If the cart accelerates too rapidly, decrease the hanging mass until you can successfully produce an acceptable acceleration rate. Measure the new hanging mass and record its value in Table 1 for Trials 2 and 3. Before performing the trial, verify that the angle of incline has not changed.

4. What would happen to the magnitude of the acceleration if the mass of the cart increased and everything else remained the same? Explain.

L For Trial 3, keep the same hanging mass that you used in Trial 2, but increase the cart's mass by taping a 500.00 g mass onto the cart's frame. Measure the total mass of the weighted cart and record this value in Table 1 for Trial 3. Before performing the trial, verify that the angle of incline has not changed.

Analysis

5. Calculate the magnitude of the expected acceleration of the cart for each trial and record these values in Table 1. Show your work for Trial 1 below. Assume no friction.

Name ____________________

M Transfer the videos to your computer.

N Use the video analysis software to analyze the three trials.

O Create position-time and velocity-time graphs for each trial.

P Determine the *experimental* average acceleration for each trial from your position-time graph or velocity-time graph and record your results in Table 1.

6. Determine the percent error between the experimental and calculated (expected) accelerations. Record your results in Table 1. Show your work for Trial 1 below.

7. What would happen to the magnitude of the acceleration if the slope of the incline increased and everything else remained the same? Explain.

____

____

____

____

Conclusion

8. How did the experimental accelerations compare to the expected (calculated) values?

____

____

9. Were the deviations of the accelerations from expected values due to errors in data collection, or was there a factor that we did not account for?

____

____

____

Going Further

10. Since friction is part of all real-world systems, there were some friction forces involved in the motion of the cart. Assume that the difference between the expected and experimental values was due to the net frictional force. Calculate the values of this net frictional force for Trial 1.

TABLE 1

Trial	Cart Mass (kg)	Hanging Mass (kg)	Incline (°)	Expected Acceleration (m/s^2)	Experimental Average Acceleration (m/s^2)	$\%_{error}$
1			20.0			
2			20.0			
3			20.0			

LAB

7A

Name

Date

Around the Curve

Investigating Circular Motion

Pilots and astronauts experience a variety of forces during flight. At times these forces can be extreme, like when pilots are executing sharp turns or ending a steep dive. Pilots and astronauts call these forces *g-forces*. The "g" in "g-force" is for gravity, and 1 g is the force caused by the acceleration due to gravity, or the pilot's normal weight. Rapidly accelerating modern aircraft and spacecraft can generate huge forces on the human body, and pilots and astronauts must be prepared for this.

Scientists have developed a training device called a *centrifuge* to prepare pilots and astronauts for the high g-forces encountered in flight. A centrifuge is a long machine that creates rotational motion by spinning on its axis. The trainee, sitting at the outside edge of the centrifuge, moves in a circle. The faster the centrifuge spins, the greater the centripetal force experienced by the rider. A pilot or astronaut is trained to be able to withstand 5 g (five times the force of gravity) without the assistance of a g-suit and up to 7 g with a g-suit. She learns how to maintain her blood pressure, keeping blood flowing to her brain. Without the skills developed during this training, g-forces would cause blood to pool in her lower body, leading to unconsciousness—and flying a plane while unconscious is very difficult to do! In this lab activity you won't experience any g-forces, but you will investigate the factors that affect circular motion.

What determines how tight a circle I can drive?

QUESTIONS

- How can I figure out how variables affect circular motion?
- What is the relationship between centripetal force and radius, mass, and tangential velocity?

EQUIPMENT

- laboratory balance
- large two-hole rubber stoppers (5)
- plastic tubing
- meter sticks (2)
- fishing line 10 lb, 1.5 m
- spring scale, 5 N
- stopwatch
- marker
- masking tape
- goggles

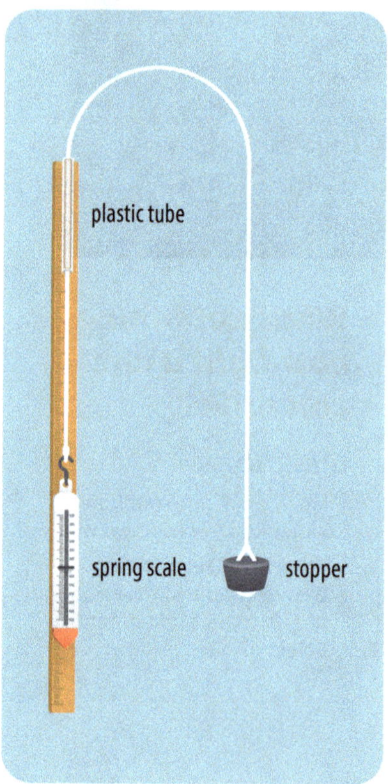

▲ Centripetal force apparatus

1. On the basis of the discussion of the centrifuge in the introduction and the image above it, what factors do you think impact the force on a pilot riding the centrifuge?

___

___

___

2. Define *centripetal force.*

___

___

Procedure

SETUP

A Label the rubber stoppers 1–5. Measure the mass of each stopper and record your data (in both grams and kilograms) in Table 1.

B Tape the plastic tube onto the end of one of the meter sticks, leaving approximately 0.5 cm of the tube extending past the end of the meter stick.

C Thread one end of the fishing line through one of the holes in Stopper 1 and back through the other hole. Tie a knot to secure the stopper. Then thread the other end of the fishing line through the tube from the top of the meter stick. Tie a loop in the free end of the fishing line.

D Lay the spring scale horizontally on a flat surface and check the zero point. Adjust the indicator scale as necessary to zero the scale. Slip the loop of the fishing line over the spring scale's hook. Your apparatus should look like the centripetal force apparatus shown at left.

E Place the entire apparatus on the floor. Pull the stopper and attached fishing line until the center of mass of the stopper is 80.0 cm from the opening of the tube. Gently pull the fishing line attached to the spring scale taut along the meter stick, maintaining the 80.0 cm to the stopper. Using the masking tape, mark the meter stick with a line at the top of the hook of the spring scale.

When performing an experimental trial, hold the meter stick and spring scale with one hand and the plastic tube and meter stick with the other. Make sure that you do not restrict the fishing line from moving. Move the hand that is holding the spring scale up and down as the spring stretches so that the hook remains lined up with the reference mark as shown at right.

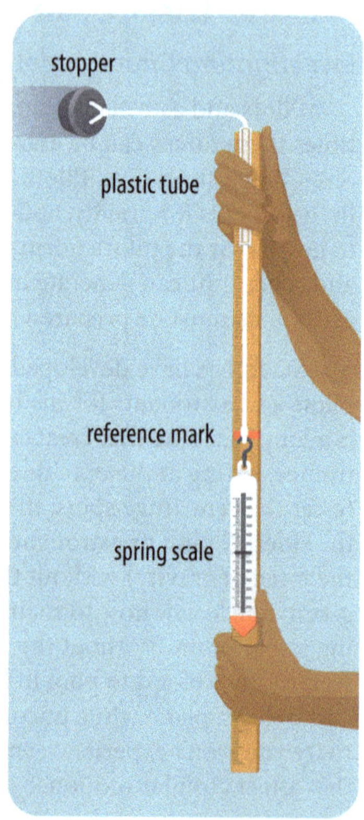

Name ______________________

DATA COLLECTION: VARYING CENTRIPETAL FORCE

F After checking the area for people or obstructions that could be struck by the swinging stopper, swing the stopper in a circular path above your head as horizontally as possible. Adjust the speed with which you swing the stopper until the scale reads 1.0 N. Once you have established the correct speed, move the spring scale up or down until the hook lines up with the reference mark.

G Once the scale is steady, direct a team member to start timing when the mass passes a reference point. Count each complete revolution of the mass out loud; stop timing when 20 revolutions are completed. Record the force, time for 20 revolutions, and stopper mass in Table 2.

H Repeat Steps F and G for four other force (1.5 N, 2.0 N, 2.5 N, and 3.0 N) and speed combinations. Be sure that the spring scale hook remains lined up with the reference mark for all measurements. Record the force and time data for each trial in Table 2.

3. What force provides the centripetal force, making the stopper move in its horizontal circle?

____

DATA COLLECTION: VARYING RADIUS

I Copy your data from Trial 3 into the Trial 8 row of Table 2.

J Place the entire apparatus on the floor. Pull the stopper and attached fishing line until the center of mass of the stopper is approximately 40.0 cm from the opening of the tube. Gently pull the fishing line attached to the spring scale taut along the meter stick, maintaining the 40.0 cm to the stopper. Using the masking tape, mark the meter stick with a line at the top of the hook of the spring scale.

K Swing the stopper in a circular path above your head as horizontally as possible. Adjust the speed with which you swing the stopper until the scale reads 2.0 N. Once you have established the correct speed, move the spring scale up or down until the hook lines up with the reference mark.

L Once the scale is steady, direct a team member to start timing when the mass passes a reference point. Count each complete revolution of the mass out loud; stop timing when 20 revolutions are completed. Record the radius and time for 20 revolutions in the Trial 6 row of Table 2.

M Repeat Steps J–L for three other radius (60.0 cm, 100.0 cm, 120.0 cm) and speed combinations. Be sure that the spring scale hook remains lined up with the reference mark for all measurements. Record the radius and time data for Trials 7, 9, and 10 in Table 2.

DATA COLLECTION: VARYING STOPPERS (MASS)

N Copy your data from Trial 5 into the Trial 11 row of Table 2.

O Attach Stoppers 1 and 2 to the apparatus.

P Place the entire apparatus on the floor. Pull the stopper and attached fishing line until the center of mass of the stopper is approximately 80.0 cm from the opening of the tube. Gently pull the fishing line attached to the spring scale taut along the meter stick, maintaining the 80.0 cm to the stopper. Using the masking tape, mark the meter stick with a line at the top of the hook of the spring scale.

Q Swing the stopper in a circular path above your head as horizontally as possible. Adjust the speed with which you swing the stopper until the scale reads 3.0 N. Once you have established the correct speed, move the spring scale up or down until the hook lines up with the reference mark.

R Once the scale is steady, direct a team member to start timing when the mass passes a reference point. Count each complete revolution of the mass out loud; stop timing when 20 revolutions are completed. Record the time for 20 revolutions and the combined mass of the stoppers in the Trial 12 row of Table 2.

S Repeat Steps O–R for each subsequent stopper being added (Stoppers 3–5). Be sure that the spring scale hook remains lined up with the reference mark for all measurements. Record the times and combined stopper masses in Trials 13–15 in Table 2.

Analysis

T For each trial, calculate the tangential velocity

$$v_t = distance/\Delta t = (revolutions)(2\pi r)/\Delta t).$$

Record the values in Table 2. Recall that Trials 3 and 8 are the same and that Trials 5 and 11 are the same. Show your work for Trial 1 below.

Name ______________________

U Create scatterplots for the following data sets from Table 2. Choose an appropriate best-fit curve for each set of data. Both best-fit curves for Graphs 1 and 2 should pass through the origin.

a. Graph 1 (Trials 1–5): F_c (*y*-axis) versus v_t (*x*-axis) (set the origin as the intercept for the curve of best fit)

b. Graph 2 (Trials 6–10): r (*y*-axis) versus v_t (*x*-axis) (set the origin as the intercept for the curve of best fit)

c. Graph 3 (Trials 11–15): m (*y*-axis) versus v_t (*x*-axis)

4. What type of curve fits each set of data best?

____

____

____

5. According to your experimental evidence (data and graphs), how would the (a) centripetal force, (b) radius, and (c) mass change if the tangential velocity of the stopper(s) were doubled and everything else remained constant?

____

____

____

6. Write a proportionality for each of the three data sets that you compared.

____

____

____

7. Discuss the error that is introduced if the stopper and fishing line do not remain horizontal. Specifically mention how the stopper's weight affects the tension on the fishing line.

____

____

____

____

____

Conclusion

8. Combine the three proportionalities from Question 6 into a single combined proportionality.

9. How well does your proportionality in Question 8 compare to the formula for centripetal force?

Going Further

10. Think back to the pilots and astronauts discussed in the opener. If a 75.2 kg pilot rides in a centrifuge with a 6.10 m radius and the centrifuge spins at 26.4 rpm, what will the pilot's tangential velocity be?

Name ______________________

11. For the pilot in Question 10, (a) what will the centripetal acceleration be? (The centripetal acceleration can be calculated with the formula shown below.) (b) How many g is this?

$$a_c = \frac{v_t^2}{r}$$

12. For the pilot in Question 10, (a) what is the magnitude of the centripetal force? (b) In what direction does this force point?

13. On the basis of what you learned in this lab activity, what do you think determines how tightly a car can turn?

____

____

____

____

____

TABLE 1 *Rubber Stoppers*

Stopper	Mass (g)	Mass (kg)
1		
2		
3		
4		
5		

TABLE 2

<table>
<tr><td colspan="8">Part 1: Varying Centripetal Force</td></tr>
<tr><th>Trial</th><th>Stopper(s)</th><th>Radius (m)</th><th>Force (N)</th><th>Revolutions</th><th>Time (s)</th><th>Mass (kg)</th><th>Tangential Velocity v_t (m/s)</th></tr>
<tr><td>1</td><td rowspan="5">1</td><td rowspan="5">0.800</td><td></td><td>20</td><td></td><td rowspan="5"></td><td></td></tr>
<tr><td>2</td><td></td><td>20</td><td></td><td></td></tr>
<tr><td>3</td><td></td><td>20</td><td></td><td></td></tr>
<tr><td>4</td><td></td><td>20</td><td></td><td></td></tr>
<tr><td>5</td><td></td><td>20</td><td></td><td></td></tr>
<tr><td colspan="8">Part 2: Varying Radius</td></tr>
<tr><td>6</td><td rowspan="5">1</td><td></td><td rowspan="5">2.0</td><td>20</td><td></td><td rowspan="5"></td><td></td></tr>
<tr><td>7</td><td></td><td>20</td><td></td><td></td></tr>
<tr><td>8</td><td></td><td>20</td><td></td><td></td></tr>
<tr><td>9</td><td></td><td>20</td><td></td><td></td></tr>
<tr><td>10</td><td></td><td>20</td><td></td><td></td></tr>
<tr><td colspan="8">Part 3: Varying Mass</td></tr>
<tr><td>11</td><td>1</td><td rowspan="5">0.800</td><td rowspan="5">3.0</td><td>20</td><td></td><td></td><td></td></tr>
<tr><td>12</td><td>1, 2</td><td>20</td><td></td><td></td><td></td></tr>
<tr><td>13</td><td>1–3</td><td>20</td><td></td><td></td><td></td></tr>
<tr><td>14</td><td>1–4</td><td>20</td><td></td><td></td><td></td></tr>
<tr><td>15</td><td>1–5</td><td>20</td><td></td><td></td><td></td></tr>
</table>

LAB

7B

Name

Date

Take It for a Spin

Exploring Angular Acceleration

The army of Israel was face-to-face with the army of the Philistines. Every day, for many days in a row, the Philistine champion, a giant named Goliath, taunted and challenged the army of Israel. Finally, a shepherd boy named David volunteered to face the giant. We all know how the story ends, but how does a boy with a sling and a stone kill a giant? A sling is a rotational system that is used as a weapon. A stone, the projectile, is held in the sling, which moves the stone in a circular path. Increasing the sling's rate of rotation increases the velocity of the stone. David was able to accelerate the stone to a speed that could kill Goliath.

The story of David and Goliath clearly demonstrates the connection between the rotational motion (sling) and the circular motion of an object (stone) within a system. The dimensions of angular motion ($\Delta\boldsymbol{\theta}$, $\boldsymbol{\omega}$, $\boldsymbol{\alpha}$) are all the same within a system of rotational motion, but the dimensions of circular motion ($\mathbf{a}_c$, $\mathbf{v}_t$) associated with that motion depends on the position (radius) involved. This becomes clear when we think about pushing on a wheel to make it rotate. Pushing near the center of a wheel causes a small angular acceleration, while pushing near the edge of the wheel causes a larger angular acceleration. For a given force applied at a certain position, the whole system will have the same angular dimensions ($\Delta\boldsymbol{\theta}$, $\boldsymbol{\omega}$, $\boldsymbol{\alpha}$).

Recall that the magnitude of centripetal acceleration is usually expressed as

$$a_c = \frac{v_t^2}{r}.$$

How are rotational motion and circular motion related?

QUESTIONS

- » What is the relationship between angular velocity and centripetal acceleration?
- » What is the relationship between radius and centripetal acceleration?

EQUIPMENT

- smartphone with physics app
- computer with spreadsheet program
- board, 15 cm × 1.0 m
- clamps or tape
- spinning stool

It can be expressed in terms of angular velocity if we consider the relation $v_t = r\omega$. We can insert this into the centripetal acceleration equation.

$$a_c = \frac{v_t^2}{r}$$

$$a_c = \frac{(r\omega)^2}{r}$$

$$a_c = \frac{r^{\cancel{2}}\omega^2}{\cancel{r}}$$

$$a_c = r\omega^2$$

We will use a simple rotational system to understand this relationship between centripetal acceleration (circular motion) and angular velocity (rotational motion).

Procedure

DATA COLLECTION

A Clamp one end of the board to the stool as shown in the image below. Position the stool so that the board will be free to rotate through 360°.

B Launch the physics app on your smartphone.

C Select the centripetal acceleration feature.

D Secure the smartphone to the board using clamps or tape.

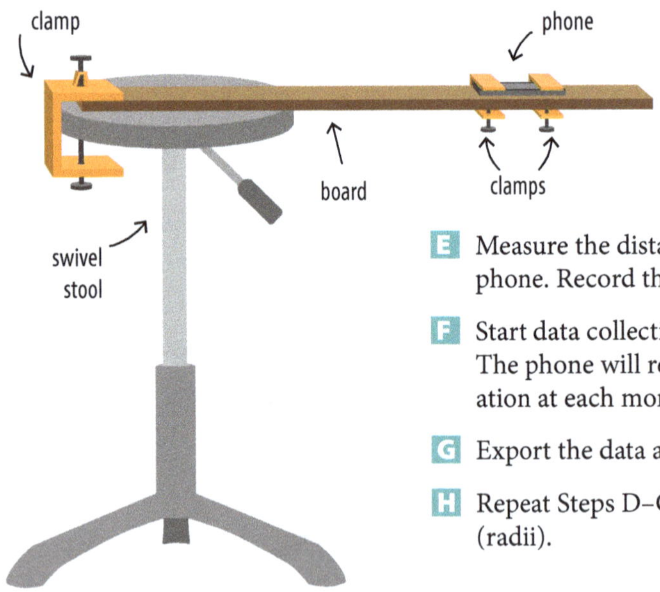

E Measure the distance from the center of rotation to the center of the phone. Record this radius in Table 1.

F Start data collection and then rotate the system at many different speeds. The phone will record the angular velocity and the centripetal acceleration at each moment.

G Export the data and send it to your computer.

H Repeat Steps D–G seven times for a total of eight different positions (radii).

Analysis

I Create a scatterplot of centripetal acceleration versus the angular velocity for each data set. Fit a second-degree polynomial to the data.

1. How well does a second-degree polynomial describe the data?

____

____

____

Name ______________________

2. What is the physical significance of your answer to Question 1?

J Record in Table 1 the coefficient of the squared term for each of the curves of best fit for the scatterplots that you created in Step I.

K Create a scatterplot of the coefficients versus radii from Table 1. Use a linear best fit for the data.

3. How well does a line describe the data?

4. What is the equation of the line of best fit? What does this tell us about the relationship between the radius and coefficient?

Conclusion

5. How is the centripetal acceleration related to angular velocity?

6. How is the radius at which the phone is spun related to its angular acceleration?

Going Further

7. Think back to David and Goliath. What factors would have affected the speed at which David slung the stone?

8. Assuming that David needed to sling the stone at a speed of 28.2 m/s to kill Goliath, use the equations mentioned above to determine (a) the angular velocity that he needed to rotate the sling and (b) the centripetal acceleration of the stone. Assume that the sling was 1.15 m long.

TABLE 1 *Coefficients at Each Radius*

Radius (m)	Coefficient

LAB
8A

Name ______________________

Date ______________________

Losing Energy?

Investigating Conservation of Energy

The climbs, dives, and high speeds make roller coasters exciting. So what goes into designing a roller coaster? One key concept that designers must understand is conservation of energy. Most roller coaster rides start with a climb to the top of a huge hill. To get there, a motor must do work on the car. At the top of the hill the car has a little kinetic energy, but mostly gravitational potential energy. The rest of the ride is a matter of conserving that energy as it is converted back and forth between potential energy and kinetic energy. While doing that, the designer must account for energy losses during the ride. It certainly would be embarrassing if the car didn't make it all the way through the ride! In this lab activity you will analyze the motion of three objects undergoing projectile motion to see whether mechanical energy is conserved.

1. Describe how you think the gravitational potential energy and kinetic energy will relate to each other as a golf ball is tossed in projectile motion.

____

____

____

____

____

____

____

Where does "lost" energy go?

QUESTIONS

» How do kinetic, potential, and mechanical energy relate to each other?

» How can I determine whether mechanical energy is conserved?

EQUIPMENT

- laboratory balance
- video camera
- computer with video analysis and spread-sheet programs
- golf ball
- tennis ball
- badminton birdie
- meter stick

2. Make a prediction about the conservation of mechanical energy for each of the three objects.

Procedure

COLLECTING THE DATA

A Measure the mass of the golf ball, tennis ball, and birdie. Record your data in your spreadsheet program.

B Set up the video camera to capture the motion of the golf ball, tennis ball, and birdie as they undergo projectile motion. Make sure that the meter stick is visible within the video frame.

C Start recording and toss the golf ball so that it will complete most of its projectile motion within the video frame. Consider making a few recordings to ensure that you have a good set of data.

D Repeat Step C with both the tennis ball and badminton birdie.

Analysis

E Use the video analysis software to analyze one movie clip each for the golf ball, tennis ball, and badminton birdie.

F Copy your time, position, and velocity data from the video analysis software to your spreadsheet for further analysis. In the spreadsheet program calculate the gravitational potential energy, kinetic energy, and mechanical energy for each object.

G Create a scatterplot for each object that includes the gravitational potential, kinetic, and mechanical energies with respect to time.

3. Describe the graphs of gravitational potential energy with time.

4. Describe the graphs of kinetic energy with time.

Name ____________________

5. Describe the graphs of mechanical energy with time.

6. Are the graphs of mechanical energy with time consistent with conservation of mechanical energy? Explain your answer.

Conclusion

7. If mechanical energy is not conserved for any of the objects, explain where the energy went.

Going Further

8. Think back to the roller coaster discussed in the opener. What are the physics principles that the roller coaster designer must consider?

LAB
8B

Name

Date

Big Bill's Better Bungees

Designing a Better Bungee

The sport of bungee jumping was introduced in the late 1970s. A jump is typically made from a structure, such as a bridge, tower, or dam. Attached to the structure with a bungee cord, a jumper leaps off the platform and experiences free fall. He falls until the bungee cord stops his descent and propels him back upward. The jumper springs up and down until finally coming to rest. Since the 1970s millions of successful jumps have occurred. While accidents have been known to occur, the sport actually has a high degree of safety because of the meticulous calculations done before a jump is made and the number of inspections and safety checks before each jump.

Whether you are willing to jump or not, the physics of bungee jumping is fascinating. The interaction of gravitational potential energy, elastic potential energy, and kinetic energy is a great study in energy conservation. The forces involved (elastic and gravitational primarily) are of interest to the physicist and exciting for the jumper. The complex interaction of speed, force, and energy make for a complex calculation that some consider as dizzying as the jump itself.

In this STEM activity you will open a bungee jumping operation for action figures. Before your first customer jumps, you will need to determine the proper length of the bungee to get your jumper as close as possible to the ground, without allowing him to hit the floor.

How can I design the best bungee jump?

QUESTIONS

- How do gravitational potential, elastic potential, and kinetic energies interact during a bungee jump?
- How can I compute the energies associated with bungee jumping?
- According to my data, how well did I design my bungee jump?

EQUIPMENT

- laboratory balance
- tape measure
- action figure
- elastic bands
- goggles

Procedure

DESIGN PARAMETERS

As you design your bungee, you must work within the design parameters listed below.

- The location of the bungee jump will determine the distance from the platform to the ground.
- Your selected action figure will introduce constraints for the height and mass of your jumper.
- The elastic bands selected will determine how the elastic characteristics will impact your bungee jump.
- Your teacher will specify how you are permitted to test your material and your bungee.
- Your teacher will specify the number of trials and modifications allowed.
- Your teacher will specify the scoring for your jump.

PLANNING THE DESIGN

A Research the physics that is involved in bungee jumping. Brainstorm how each of the physics principles will impact the design of your bungee. Some questions to consider: What happens to the action figure's energy during the jump? Where is your zero for potential energy? Where on your action figure will you measure when calculating potential energies?

B Keeping in mind the design parameters listed above, brainstorm possible designs for your bungee with your group.

C You will need to conduct some testing to determine the elastic properties of your bungee cord. Questions to think about: How can you determine the elastic (spring) constant for the bungee cord? Does this constant remain the same as you combine elastic bands? In other words, is the constant a constant for a particular bungee (group of elastic bands) or for the material?

D Develop a plan for your bungee cord. Indicate on this plan that this is a *preliminary design*. Some questions to consider: Where do you want your action figure to stop? What are the energies (gravitational potential, elastic potential, and kinetic) at the top? What are the energies at the bottom? What displacement will result in a successful bungee jump? What length of cord do you need for a successful bungee jump?

E Share your research findings with your group. Reach consensus on a final design that your group will build and then test it.

F Have a group member develop a plan for the group's bungee cord. Indicate on the diagram that this is an *initial production design*.

Name ____________________

BUILDING THE BUNGEE

G Using the drawing created in Step F, construct your bungee cord.

TESTING THE DESIGN

H Conduct your first trial with your action figure. Videotaping the jump will allow you to determine the clearance from the ground. Video analysis software will allow you to make accurate measurements.

REFINING THE DESIGN

I Evaluate the performance of your bungee cord.

J Make changes to your cord, if desired. If you decide to make modifications to your design, produce a *modified production design*.

RETESTING THE DESIGN

K Take your action figure up for another attempt if you made modifications.

LAB

9A

Name

Date

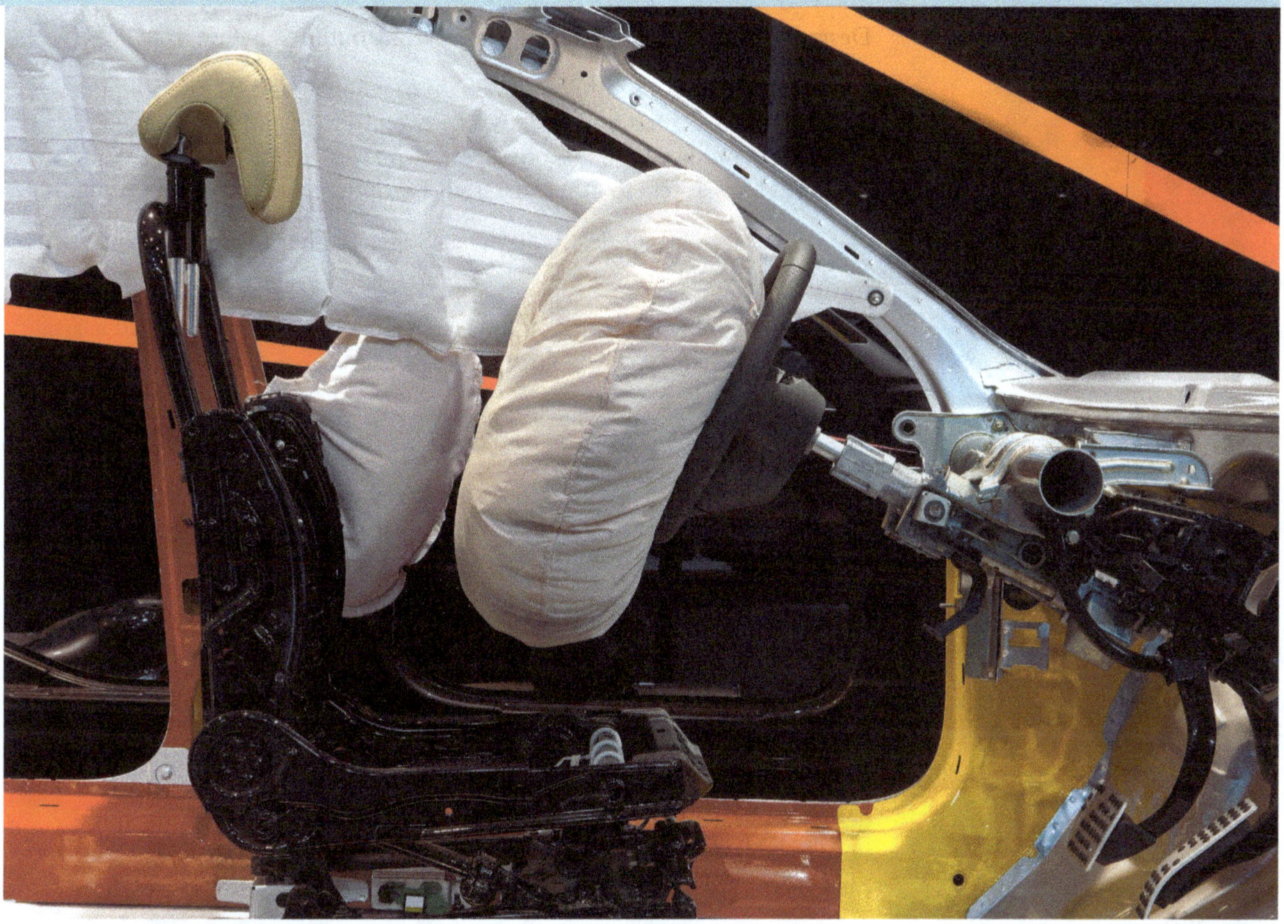

Striving for Eggcellence

Designing a Safer Restraint System

Around six million car accidents are reported each year in the United States, and the National Highway Traffic Safety Administration estimates that another ten million unreported accidents occur each year. While there are many causes of accidents and experts work to address them, it seems inevitable that as long as we drive cars, we will have accidents. And at what cost? There are millions of injuries and thousands of deaths and billions of dollars in monetary damages.

If accidents are inevitable, how can we save lives and reduce costs? Over time manufacturers have added brakes, seat belts, and airbags, and then worked to improve those systems. The most recent improvement includes additional airbags to protect occupants from side impacts. In this activity you will design, build, and test a car that will protect its passengers from the forces involved in a collision.

How can I design a safer vehicle restraint system?

QUESTIONS

- How do I design a car that provides for the safety of the occupants?
- How can I maximize the speed and safety of a car?
- What does my data tell me about how well I designed my car?

Procedure

DESIGN PARAMETERS

Design and build a car within the constraints outlined below.

- Materials for both the final version of your car and its safety equipment are limited to the following: two pieces of computer paper, 30 cm of masking tape, one straw, two axles, and four wheels. You will have access to other materials for doing developmental testing.
- Your teacher will show you the testing apparatus and explain the test procedures.
- Your score will depend on how fast your car goes and how well it protects its passenger—an egg.

PLANNING THE DESIGN

A Research the physics involved in automotive safety equipment. Consider an internet search using the keywords "airbags," "automotive passenger safety," "automotive passenger restraint," and "SAFER barrier."

B Brainstorm with your lab group how you will use the limited resources to build a car and provide for the safety of the egg.

C You will need to conduct some testing to determine the best design for both the car and its passenger protection system.

D Each member of the group should draw a sketch of his recommended design. Indicate on this plan that this is a *preliminary design*. Some questions to consider: How can we get the car to go as fast as possible? How can we best protect the egg? How do we balance speed with safety?

E Discuss with your group members the positive and negative aspects of each design and come to a consensus on the best design.

F Have someone in the group draw a diagram of the group's planned car design. Indicate on the diagram that this is an *initial production design*.

Name

BUILDING THE DESIGN

G Construct your car according to the drawing created in Step F.

TESTING THE DESIGN

H Conduct a series of tests on your vehicle at a variety of increasing speeds. Record appropriate data so that you can determine the car's speed, change in momentum, kinetic energy, and impulse. Video recording the trials is a great technique for collecting this data. You can then do video analysis of your trials. The videos will also be fun to watch after testing is done.

REFINING THE DESIGN

I Evaluate the performance of your car and recommend modifications to the design that you think will improve its performance.

J Make a drawing of your improved design. Indicate on the drawing that this is the *modified production design*.

K Build your modified car if time permits.

RETESTING THE DESIGN

L Test your modified design and evaluate the effectiveness of its modifications.

LAB
9B

Name ____________________

Date ____________________

Collision Course

Investigating Conservation during Collisions

The first time someone shoots a rifle or shotgun, he gets a quick lesson in conservation of momentum. When a gun is fired, the force of the exploding propellant imparts momentum to the bullet. At the same time, the propellant gases also impart an equal amount of momentum to the rifle. To simplify this system, we can neglect the propellant gases and assume that the bullet and the rifle exert forces directly on each other. Newton's third law states that if the rifle exerts a force on the bullet, the bullet must exert an equal but opposite force on the rifle. The result of this force is called the *kick*, or *recoil*, of the rifle.

The principle of the conservation of momentum states that the sum of the changes of momentums of the rifle and bullet during the firing event must equal zero. In this lab activity we won't be shooting any guns, but we will investigate systems during collisions to observe what is conserved during these collisions.

1. Define *momentum*.

____

____

2. Write a formula that expresses the law of the conservation of momentum.

____

What remains constant during a collision?

QUESTIONS

» How can I analyze the complex interactions during a collision?

» What should the data look like when momentum and kinetic energy are conserved during a collision?

EQUIPMENT

- video camera
- laboratory balance
- computer with video analysis and spreadsheet software
- metric ruler
- dynamics carts (2)
- metric mass set
- Velcro® tape

3. A 11.0 g bullet leaves the muzzle of a 3.25 kg rifle with a muzzle velocity of 815 m/s. What is the change in momentum of the bullet?

4. If the rifle were free to move when the bullet is fired, what would the velocity of the rifle be as the bullet left the barrel?

Name ______________________

Procedure

A Set up the frame for your videos. You will need a level area (your carts should roll freely but should not roll on their own) and about 2.0 m of working space. Set the metric ruler so that it will be in the video frame parallel to the carts' movements.

B Prepare your carts. Depending on the style of dynamics cart that you are using, you may need to attach a flexible metal spring strip to Cart 1. Attach a piece of Velcro hook tape to the end of Cart 1 that does not have the spring strip or plunger. Attach a piece of Velcro loop tape to one end of Cart 2.

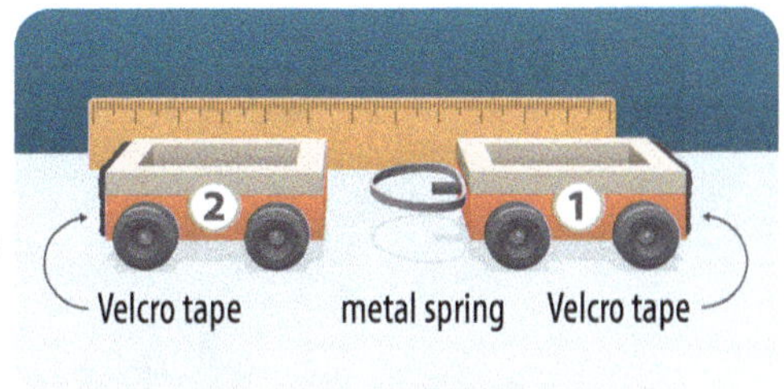

C Label both sides of Carts 1 and 2 so that the numbers will be visible in your videos. Measure the mass of each cart and record the values in Table 1.

FIRST COLLISION MODE

D Place Cart 2 in the center of the video frame with its Velcro tape facing away from the direction from which you plan to roll Cart 1.

5. If both carts had the same mass, what would you expect to be true about their velocities before and after the collision?

____

____

E Trial 1: Arrange Cart 1 properly. If the cart has a metal spring strip, face the strip toward Cart 2. If your cart is equipped with a plunger, face the plunger end of Cart 1 toward Cart 2. Start recording and then roll Cart 1 from one side of the frame so that it collides with Cart 2.

F Add sufficient mass to Cart 1 to approximately double its mass and record the new total mass for Trial 2. Repeat Step E for Trial 2.

G Again, double the mass of Cart 1 and record the new total mass for Trial 3. Repeat Step E for Trial 3.

SECOND COLLISION MODE

H Copy the empty cart masses from Trial 1 to Trial 4. Sum the cart masses and enter as the final mass for Trial 4.

I Place Cart 2 in the center of the video frame with its Velcro tape facing toward the direction from which you plan to roll Cart 1.

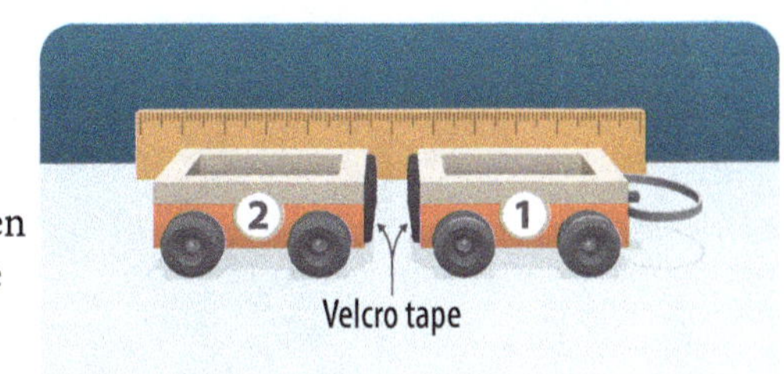

J With the Velcro tape end of Cart 1 facing Cart 2, start recording and then roll Cart 1 from one side of the frame so that it collides with Cart 2. The carts should latch onto each other.

K Add sufficient mass to Cart 1 to approximately double its mass and record the new masses for Trial 5. Repeat Step J for Trial 5.

L Again, double the mass of Cart 1 and record the new mass for Trial 6. Repeat Step J for Trial 6.

Analysis

M Use the video analysis software to analyze the motion of each cart for each trial before and after the collision. Record the velocity of each cart before and after the collision in Table 1.

N Calculate the initial and final momentums for each cart for each trial. Record the momentums in Table 2. Show your work for Trial 1 below.

O Calculate the final momentum of the system for each trial and record your calculation in Table 2.

P Calculate the percent difference between the system momentums for each trial and record your calculations in Table 2. Show your work for Trial 1 below.

Name

Q Calculate the initial and final kinetic energies for each cart for each trial and record your calculations in Table 2. Show your work for Trial 1 below.

R Calculate the initial and final kinetic energies of the system for each trial and record your calculations in Table 2.

S Calculate the percent difference between the system kinetic energies for each trial and record your calculations in Table 2. Show your work for Trial 1 below.

6. How did the final speeds change from Trial 1 to Trial 3 as mass was added to Cart 1? Explain.

Conclusion

7. Did all six collisions demonstrate conservation of momentum? Explain.

8. Did all six collisions demonstrate conservation of energy? Explain.

9. Which collisions were closer to being perfectly elastic? Explain.

Going Further

10. In the practice problems at the start of the lab activity the bullet ended up with a high velocity, while the rifle had a relatively low velocity. Explain the reason for this.

Name

11. When firing a rifle, the rifle does not fly backward at 3–4 m/s. Why is this? Explain how this differs from what you did in this activity.

TABLE 1 *Collision Data*

First Collision Mode

Trial	$m_{Cart\ 1}$ (kg)	$m_{Cart\ 2}$ (kg)	Initial v_1 (m/s)	Final v_1 (m/s)	Final v_2 (m/s)
Empty					
1					
2					
3					

Second Collision Mode

Trial	$m_{Cart\ 1}$ (kg)	$m_{Cart\ 2}$ (kg)	Initial v_1 (m/s)	Final m (kg)	Final v_2 (m/s)
4					
5					
6					

TABLE 2 *Calculations*

Momentum

First Collision Mode

Trial	Initial p_{system} (kg·m/s)	Final p_1 (kg·m/s)	Final p_2 (kg·m/s)	Final p_{system} (kg·m/s)	Percent Difference
1					
2					
3					

Second Collision Mode

Trial	Initial p_{system} (kg·m/s)	Final p_{system} (kg·m/s)	Percent Difference
4			
5			
6			

Kinetic Energy

First Collision Mode

Trial	Initial	Final			Percent Difference
	K_{system} (kg·m/s)	K_1 (kg·m/s)	K_2 (kg·m/s)	K_{system} (kg·m/s)	
1					
2					
3					

Second Collision Mode

Trial	Initial	Final	Percent Difference
	K_{system} (J)	K_{system} (J)	
4			
5			
6			

LAB
10A

Name ____________________

Date ____________________

Taking a Swing at It

Inquiring into the Period of a Pendulum

There is something soothing about a rocking motion. Babies love to be rocked to sleep. Children love the rhythmic swing of a playground swing. Many people love to rock in a rocking chair on the front porch. These are all examples of periodic motion with a repeated motion. It is this consistently repeated motion that powers grandfather clocks. The mass swings back and forth, keeping time. Did the clock designer need to know how long to make the arm of that pendulum? Does the mass of the pendulum matter? Is there a specific displacement that you should use when starting a grandfather clock? In this inquiry activity you will design an experiment to determine what variables affect the period of a pendulum.

What determines the period of a pendulum?

QUESTIONS

- How can I design an experiment that will determine the variables that affect the period of a pendulum?
- How can I use scatterplots to determine the relationship between variables?
- How can I use curve fitting to determine the formula for the period of a pendulum?
- How do I know whether my experimental design is effective?

Procedure

PLANNING/WRITING SCIENTIFIC QUESTIONS

A Brainstorm with your lab group the variables that may affect the period of a pendulum.

B Write hypotheses for how each of the variables that you identified in Step A will affect the period of the pendulum.

C Write a series of scientific questions that will guide your experimental design to test your hypotheses from Step B.

DESIGNING SCIENTIFIC INVESTIGATIONS

D Work with your lab group to determine how you can test each of the variables independently while controlling the other variables.

E Write your procedures for testing each of the variables that you identified in Step A.

F Have your teacher approve your procedures.

CONDUCTING SCIENTIFIC INVESTIGATIONS

G Conduct your investigation according to your procedures. Keep in mind the need to control all variables except for the independent and dependent variables.

DEVELOPING MODELS

H Create scatterplots of the collected data.

I Use curve fitting to determine the relationship between each variable and period of the pendulum.

J Make a formula for the period of a pendulum from the relationships that you determined in Step I.

SCIENTIFIC ARGUMENTATION

K State a claim about each of your hypotheses from Step B. As always, support your claims with specific evidence from your data or analysis.

L State a claim about the effectiveness of your experimental design. Support your claims with specific evidence from your data or analysis.

M How could you improve the design of your experiment to produce more accurate results?

Going Further

N Look up the formula for the period of a pendulum. Can you determine the relationship between the coefficients from your scatterplot(s) and the formula?

LAB

10B

Name

Date

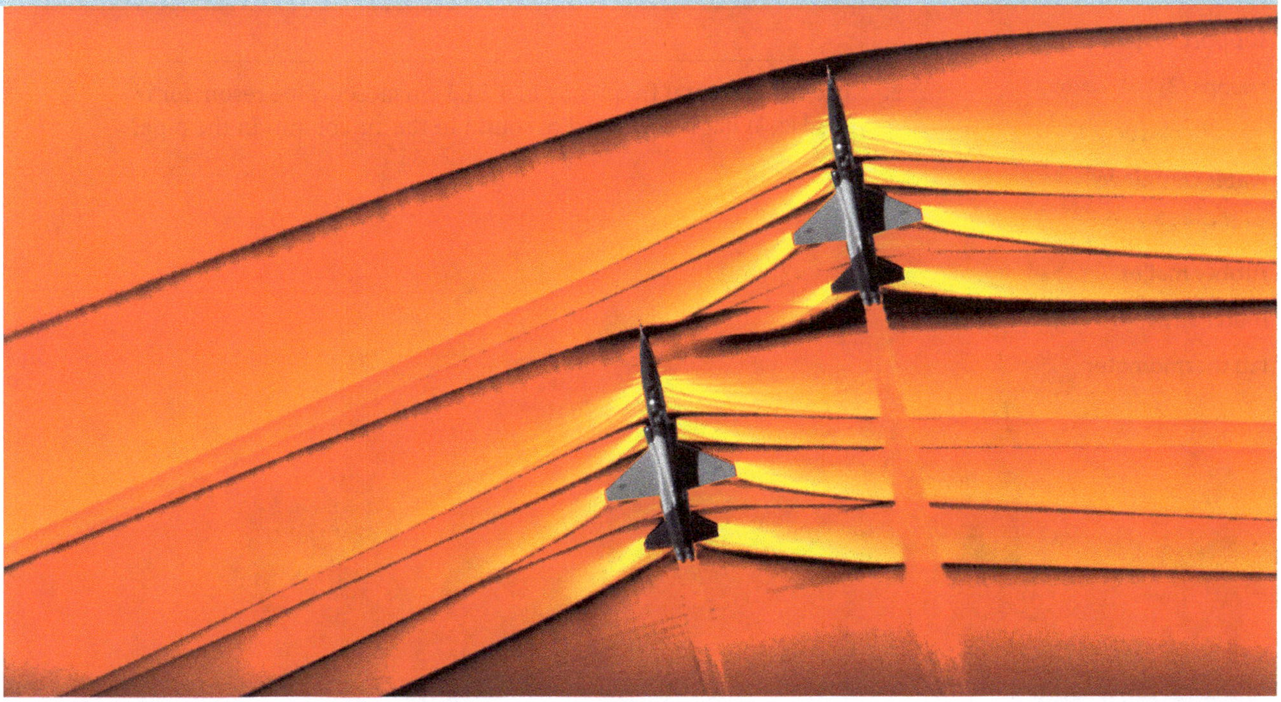

Sonic Boom

Measuring the Speed of Sound in Air

One common experience that we may have noticed is that we often hear a vehicle approaching before it passes us. This happens because the sound of the vehicle moves out ahead of it. If you have been to an airshow or have ever lived near an Air Force base, you may have experienced high-performance aircraft passing by and then hearing the sound. That sound is often in the form of a huge boom—a sonic boom. This occurs because the aircraft is traveling faster than the speed of sound. A sonic boom is the accumulation of multiple compressions of a sound wave that all reach our ears simultaneously. Sound moves really fast, so how fast must that plane be moving to be getting ahead of its sound? In this lab activity you will use a closed-pipe resonator to measure the speed of sound in air.

1. Draw a sound wave that represents the fundamental frequency in a closed pipe.

How fast do sound waves move?

QUESTIONS

» How can I measure the wavelength of a sound wave?

» What relationship is there between the frequency of a sound, the speed of sound, and the length of an air column that will cause resonance?

» How accurately can we measure the speed of sound in the laboratory?

EQUIPMENT

- resonance tube
- support system
- clamps (3)
- flexible plastic tube
- water container
- tuning forks (3)
- rubber mallet
- meter stick
- lab thermometer

2. How does the wavelength of the fundamental frequency wave compare to the length of a closed-pipe resonator?

3. The fundamental frequency in a 34.3 cm closed-pipe resonator is 256 Hz. What is the speed of sound in the air column in the pipe?

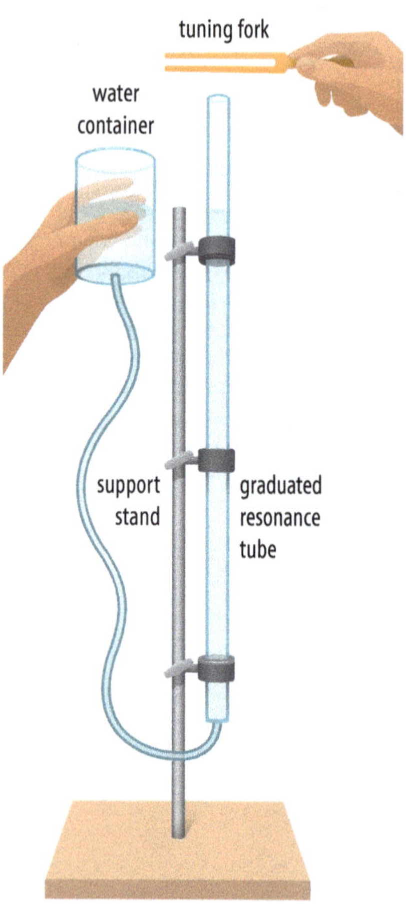

Acoustic resonance apparatus

Procedure

SETUP

A Set up the resonance apparatus as shown in the image at left. Check that the stopper in the bottom of the tube is watertight.

B Hold the water container at the level of the bottom of the resonance tube. Add water to the water container until it is about three-quarters full. This should allow you to fill the resonance tube by lifting the water container to the top of the tube. If you cannot fill the resonance tube this way, add more water to the water container.

4. How is the resonance tube in the apparatus a closed-pipe resonator? How would the setup have to be different to be an open-pipe resonator?

C Strike one of the tuning forks with a rubber mallet. If you don't have a rubber mallet, you may strike the fork on your knee or the sole of your shoe. ***Do not strike a tuning fork on a tabletop or other hard surface as damage to the tuning fork may result.*** Hold the vibrating tuning fork so that it is horizontal and as close as possible to the mouth of the tube. Do not touch the resonance tube with the tuning fork or the tuning fork will stop vibrating.

D While you hold the tuning fork near the mouth of the resonance tube, have a lab partner lower the water container to lower the level of the water in the resonance tube (increasing the length of the air column in the tube). Listen for a marked increase in the sound intensity, which indicates the resonance point. The resonance point can be more accurately determined by raising and lowering the water container a few centimeters above and below the resonance point several times. Continue to lower the water container and see whether you can find any additional resonance points.

5. Why does the sound get louder when you reach a resonance point?

____

____

____

E Measure the length of the air column (closed-pipe resonator) in the resonance tube.

F Once you are comfortable with the operation of the apparatus, continue to Step G to begin collecting data.

DATA COLLECTION

G Measure the temperature of the air in the resonance tube and record it in Table 1.

H Measure the inner diameter of the resonance tube and record the value in Table 1.

I Raise the water container until the water level in the resonance tube is nearly to the top.

J Note the frequency stamped on the first tuning fork to be used and record this value in Table 1 for each trial in which it is used.

K Strike the tuning fork with a rubber mallet. ***Do not strike a tuning fork on a tabletop or other hard surface as damage to the tuning fork may result.*** Hold the vibrating tuning fork so that it is horizontal and as close as possible to the mouth of the tube.

L Lower the water container to find the resonance point. Have a partner measure the air column length and record this measurement in Table 1.

M Continue to lower the water container to the next resonance point. Measure and record the air column length in Table 1.

N Repeat Step M to find any additional resonance points. Each resonance point should be recorded as a new trial in Table 1.

O Repeat Steps I–N for two additional tuning forks.

Analysis

The resonance point is affected by a phenomenon called *end effect*. The length of the wave formed in a closed pipe is slightly longer than the air column in the pipe. The total length is $L = l + 0.4d$, where l is the length of the resonance point and d is the diameter of the tube.

P For each air column length, make the end correction using the end correction equation to find the corrected air column length. Record these values in Table 1. Show your work for Trial 1 below.

Q Calculate the wavelength of the sound wave by multiplying the corrected air column length by the appropriate factor, depending on the resonance length. For the shortest length ($\lambda/4$) multiply by 4/1, for the next position ($3\lambda/4$) multiply by 4/3, and so on to obtain the value for λ and record these values in Table 1. Show your work for Trial 1 below.

R Use the tuning fork frequency and the wavelength calculated in Step Q to compute the experimental speed of sound for each trial and record these values in Table 1. Assume that the frequency is an exact value. Show your work for Trial 1 below.

Name ______________________

The speed of sound in air depends on the air temperature. The speed of sound at 0 °C (v_0) is 331.3 m/s. The formula for the expected speed of sound at other temperatures is

$$v_T = v_0 + (0.606\ \text{m/s·°C})T_C.$$

S Use the equation above to calculate the expected speed of sound at the temperature that you recorded in Table 1. Show your work below and record this value in Table 1.

6. Why do you think the speed of sound in air depends on what you learned today? Explain.

____

____

____

____

____

____

T Compute the percent error for each trial and record your results in Table 1. Show your work for Trial 1 below.

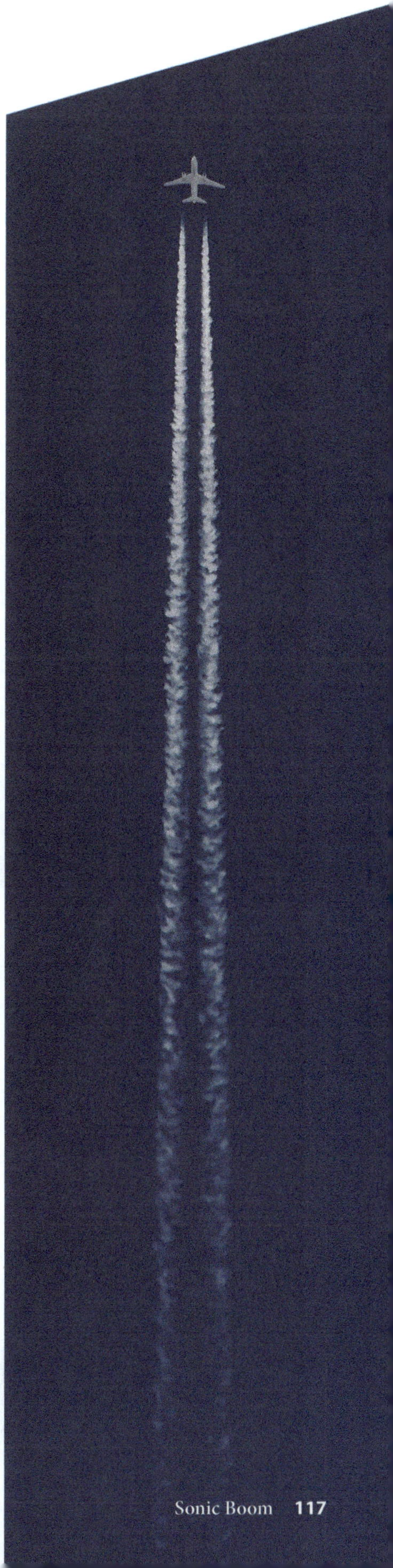

U Determine the average of your experimental speeds of sound. Determine the average percent error from this result. Record your values in Table 1.

7. If you were not aware of end correction and just assumed that the air column length of the resonance tube was $\lambda/4$, which frequency range would produce a greater error in determining wavelength: higher frequency or lower frequency, assuming that you use the same tube? Explain your answer.

Conclusion

8. Once you have determined the speed of sound for your laboratory conditions, explain how you could find the frequency of an unlabeled tuning fork.

9. Considering the experimental apparatus and procedures, what two factors do you think probably contributed the most to errors in the results?

10. What other factors do you think would affect the speed of sound in air?

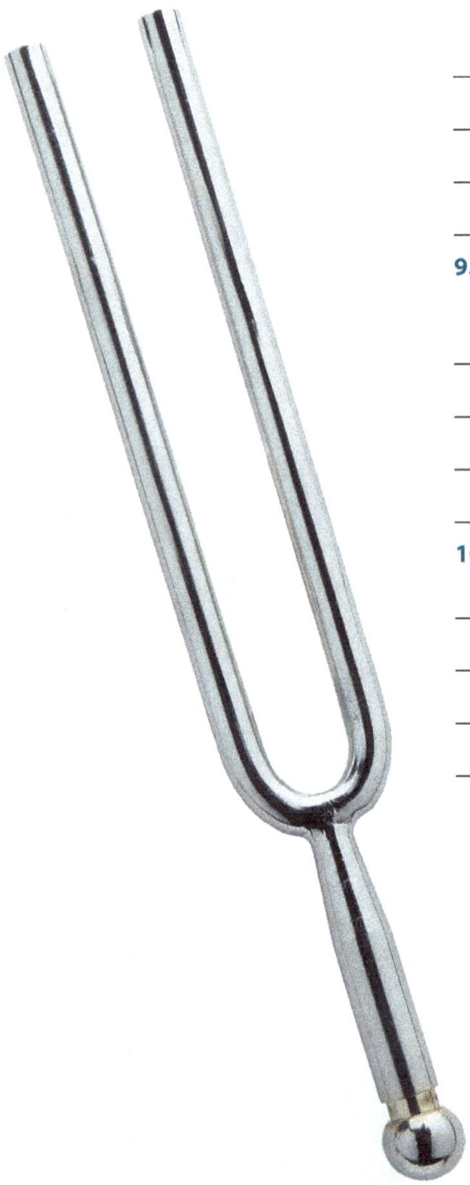

Name ______________________

Going Further

11. How could you change the experimental procedures to test the temperature dependence of the speed of sound? What challenges would you expect to have with such a procedure?

12. On the basis of what you learned today, do you think that an aircraft must go faster at a low altitude or a high altitude to break the sound barrier? Explain.

TABLE 1 *Speed of Sound Data and Calculations*

Room Temperature (°C)				Tube Diameter (m)			
Trial	**Frequency (Hz)**	**Air Column (m)**	**Corrected Air Column (m)**	**Wavelength (m)**	**Experimental Speed of Sound (m/s)**	**Speed of Sound (m/s)**	**Percent Error**
1							
2							
3							
4							
5							
6							
7							
Avg.							

LAB
11A

Name ______________________

Date ______________________

Across the Expanse

Investigating Thermal Expansion

It was early morning in July 1996 as the construction crew continued working on the 16 mi stretch of highway through Halawa Valley in Hawaii. Though it was going to be hot, the crew expected the day to go like any other. Little did the workers know that the day would be anything but routine. As the temperature soared, one of four girders that they were working on buckled. The failure of the 120 ft girder caused the three girders running parallel to it to also fail. All four girders collapsed, injuring several workers.

All materials expand and contract as they change temperature. Engineers must take these changes into account when designing structures. Buildings, bridges, highways, and train tracks are just a few of the structures that must have design elements that allow for expansion of their materials. In this lab activity you will use a linear expansion apparatus to measure the thermal expansion of different metals. You will then determine the coefficient of thermal expansion for each metal and compare your experimental values to the accepted values.

1. What factors determine how much a material will change in length with temperature change?

____

____

How much do metals expand in response to temperature changes?

QUESTIONS

» Do all materials expand equally as temperature changes?

» How can I determine the coefficient of thermal expansion?

EQUIPMENT

- steam generator and heat source
- tube-type linear expansion apparatus
- metal sample tubes
- meter stick
- thermocouple and multimeter
- beaker, 250 mL
- goggles
- hot mitts (1 pair)

2. A bar of steel is heated 50 °C. How will the change in length of a 1 m bar compare with the change in a 10 m bar?

3. What is the symbol for the coefficient of linear expansion? Starting from the equation for the change in length formula, solve for the coefficient of linear expansion.

4. A steel tube is measured and found to be 695.0 mm long. Its initial temperature is 24.5 °C. The tube is heated, and the length is changed by 0.83 mm. The tube's temperature is 97.8 °C. Determine steel's coefficient of linear expansion. Leave your answer in a form with a power of 10^{-6} like the values given in your textbook.

Name ____________________

5. If the tube in Question 4 could be heated to 200.0 °C, what would its new length be?

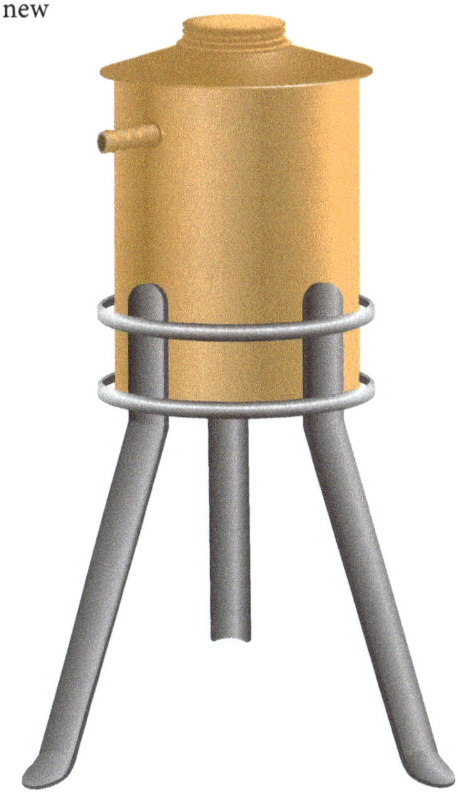

Steam generator

Procedure

A Record the kind of material for each tube to be tested (called the sample) in Table 1. Obtain the accepted coefficient of linear expansion for each metal and record it in Table 1.

B Fill the steam generator about half full of water and begin heating it. ***Do not let the steam generator boil dry!*** Do *not* connect the hose to the apparatus yet.

C Set up the apparatus for the first metal sample (see image below). Clamp or otherwise secure the tube onto the fixed end of the base. Place the support for the roller indicator under the opposite end of the tube. Measure the length of the tube between the two points of support and record it as the initial length in Table 1. Measure the temperature of the metal tube at its midpoint using a thermocouple probe attached to a multimeter. Record this value as the initial temperature in Table 1. Adjust the roller pointer to the zero mark on the scale. Attach the hose from the steam generator to the fixed end of the tube.

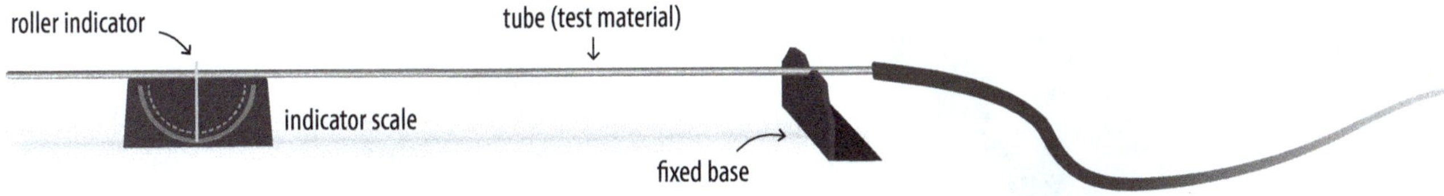

D When the apparatus begins to emit steam, start monitoring the temperature at the midpoint of the tube. When the temperature stabilizes at a maximum, allow the system to steam for 4–5 minutes more to permit the sample to achieve thermal equilibrium. Measure the temperature of the metal tube at its midpoint using a thermocouple probe attached to a multimeter and record this final temperature in Table 1.

E Note the position of the pointer on the indicator scale. Record this value as the change in length in Table 1. ***The tube and steam generator surfaces are hot!***

F Turn off the heat supply to the steam generator. Using hot mitts or towels to protect your hands, disconnect the steam supply hose from the tube. Unclamp the tube and pour out the hot condensate remaining inside the tube into a container. Take the tube to a sink and pour cool water through it until it is near room temperature.

G Repeat Steps B–F for a second trial with the same sample to evaluate the precision of the experimental technique.

H Repeat Steps B–F for the other metal tubes that you have available.

I When you have finished collecting data, carefully pour the hot water remaining in the steam generator into the sink and disassemble the apparatus according to your teacher's instructions.

Analysis

J Compute the experimental coefficient of linear thermal expansion for each metal. Leave your answer in a form with a power of 10^{-6} like the linear thermal expansion values given in your textbook. Record all the results in Table 1. Show your work for the first sample below.

Name

K Compute the percent error between your experimental values and the accepted values of the coefficient of linear expansion for each metal. Record all the results in Table 1. Compute the average percent error to evaluate the precision of the experimental methodology. Show your work for the first sample below.

L For the metal sample tested twice, compute the percent difference between the two experimental values of the coefficient of thermal expansion. Record your answer in Table 1.

6. This process assumes that the bar is uniformly heated. Why can this condition not be met with the type of apparatus used in this experiment? Explain.

7. How does the experimental procedure compensate for the concern noted in Question 6?

8. If you had measured the original length of the metal sample in inches and the apparatus indicated the change in length in inches, what would the unit α be in this case? Explain your answer.

9. Which measurement error will have a greater effect on the results of your calculations: a 1 mm error in measuring the original length or a 0.01 mm error in measuring the change in length? Prove your answer. Assume that the initial length is 600.0 mm and the change in length is 0.60 mm.

10. Do you believe that you need an instrument more precise than a meter stick to measure the metal sample's original length? Explain your answer.

Name ____________________

11. If you performed the test twice on the same sample, what does your percent difference between the two trials reveal? What does the average percent error reveal? On the basis of these values, discuss the precision and accuracy of the results that you can expect to obtain from this experiment.

____

____

____

____

____

____

Conclusion

12. Engineers need to know how materials will change shape with temperature. Using the first sample that you worked with in this experiment, calculate the length of the rod if it had cooled from room temperature to 0.0 °C. Use the accepted value α for the metal.

Going Further

13. How do engineers account for thermal expansion in their designs?

____

____

____

____

____

14. What do you think caused the girder collapse discussed in the opening paragraph?

____

____

____

____

____

TABLE 1 *Thermal Expansion Data and Calculations*

Sample	Accepted α (× 10^{-6} °C^{-1})	Initial Length (mm)	Initial Temp (°C)	Final Temp (°C)	Length Change (mm)	Experimental α (× 10^{-6} °C^{-1})	Percent Error	Percent Difference
						Average		

LAB

11B

Name ______________________

Date ______________________

Under Pressure

Investigating Boyle's Law

The invention of the airplane has made traveling much easier. In a matter of hours we can fly across an ocean and avoid a sea voyage that can take several days. We can travel across a continent and not have to sit all day while driving a car.

As airplane design has improved, planes have been able to fly higher and higher. But how do the passengers have enough oxygen? Modern aircraft are pressurized, and so an aircraft's environment might be the equivalent of 8000 feet in elevation while the aircraft is flying at 35,000 feet. This benefits everyone because aircraft that fly higher can also fly faster, saving the passengers time and the airlines money. But what happens if an aircraft loses its pressurization? The loss of cabin pressurization is one of the most critical emergencies that an aircraft crew can face. In the event of rapid decompression, the crew has seconds to put on oxygen masks before they are rendered unconscious. An aircraft's cabin pressure and the change in volume as air escapes it are related. In this lab activity you will investigate the relationship between the pressure and volume of a gas.

How can I make a balloon expand without adding air to it?

QUESTIONS

- Does the ratio of volume to pressure in a sample of gas change?
- How does the volume of a gas change in response to changes in pressure?

EQUIPMENT

- Boyle's law apparatus
- laboratory thermometer
- barometer
- metric ruler
- metric masses, 500 g (6)

1. Define *pressure*.

2. When you blow air into a balloon, you add air particles to the inside of the balloon, increasing the pressure. The higher pressure inside the balloon causes the balloon to expand. When you stop adding air, why does the higher internal pressure cause the balloon to expand only so much? What causes the balloon to stop expanding?

3. Consider the syringe of the Boyle's law apparatus shown on the facing page. If the cap is not on, so that the syringe is open to the atmosphere, what is the air pressure within the syringe? If we cap the syringe without changing any conditions of the syringe, will the pressure within the syringe change?

4. Think about what will happen to the gas in the syringe in Question 3 when we place the mass on the platform. Write a hypothesis about the effect of changing the pressure of a gas on the volume of the enclosed gas.

Name

Procedure

A Measure the temperature and air pressure in the room. Record your data in Table 1. If you don't know the current atmospheric pressure, assume that the pressure is normal atmospheric pressure: 101.325 kPa.

B Remove the plastic cap from the tip of the syringe. Remove the plunger from the syringe. Measure the inside diameter of the cylinder and record this value in Table 1.

C Reinsert the plunger and set the inner edge of the plunger at the 30 cm^3 mark. Replace the plastic cap. Do not remove the cap again until the experiment is completed. If the cap comes off at any time during the remainder of the experiment, you will have to start over.

D Gently pull up on the plunger slightly and release it. Repeat this action two or three times. Note the position of the inner edge of the plunger against the scale printed on the cylinder.

E Gently push down on the plunger and release it. Note the position of the plunger again.

F Determine the position halfway between the readings noted in Steps D and E. Record this halfway position reading as the initial volume in Table 1.

5. What do you think is the purpose for Steps D–F?

____

____

G Assemble the apparatus as illustrated in the image at right. Install the syringe in the base support block and place the platform on the top end of the syringe plunger. Balance a 500.0 g mass on the platform. You may have to support the mass to keep it from tipping laterally, but do not support any of the weight of the mass or hinder the motion of the plunger. Measure and record the volume for Trial 1 in Table 1.

H Repeat Step G for Trials 2 through 6, adding 500.0 g more mass for each trial. ***Note that as the number of masses on the platform increases, the apparatus will become more unstable without support.***

I Remove all masses. Without removing the syringe cap, perform Steps D–F again to verify the initial volume. If this volume varies from the volume obtained earlier by more than 0.5 cm^3, you should ask your teacher for assistance to determine the cause.

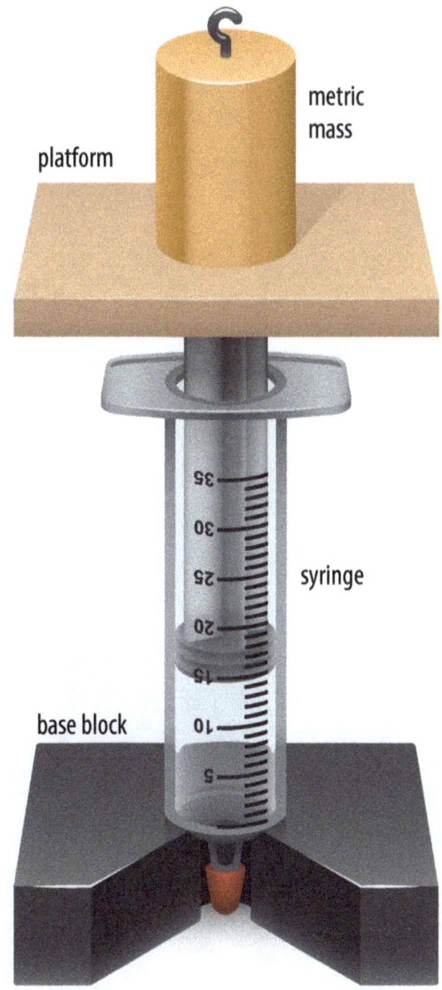

Boyle's law apparatus with a laboratory mass

Analysis

J Calculate the added force exerted by the added mass for each trial and record the values in Table 1. Show your work for Trial 1 below.

K Calculate the added pressure exerted by the added force. Record the values in Table 1. Show your work for Trial 1 below.

L Calculate the total pressure (atmospheric pressure plus added pressure) within the syringe for each trial. Record the values in Table 1.

M Create a scatterplot of volume versus total pressure. Include the origin in your scale when labeling the axes. As always, include a curve of best fit.

6. Does the scatterplot of volume and total pressure appear to have a linear best fit? Describe the graph.

____

____

____

N Create a scatterplot of volume versus the inverse of total pressure. Again, include a curve of best fit.

7. Does the scatterplot of volume and the inverse of total pressure appear to have a linear best fit? Describe the graph.

____

____

____

O Determine the slope of the volume and the inverse of the total pressure graph. This is the experimental value of the Boyle's law constant for your sample of gas. Record this value in Table 1.

Name ______________________

8. Calculate the expected Boyle's law constant using the formula

$$k = 10.299 \frac{\text{cm}^3 \cdot \text{kPa}}{\text{K}} T_K,$$

where T_K is the Kelvin temperature of your gas (assume that the gas was at room temperature). Record this value in Table 1.

9. Calculate the percent difference between your experimental and expected values for the Boyle's law constant. Record this value in Table 1.

10. State three specific sources of potential experimental error in this exercise.

____

____

____

____

Conclusion

11. On the basis of your graphical results, how would you describe the relationship between the volume and pressure of a gas in an enclosed container?

____

____

____

12. State a claim about your hypothesis. Support your claim with evidence from your data or analysis.

____

____

____

Going Further

13. The ratio of oxygen to other gases is relatively constant throughout most of the atmosphere. Why do you think that aircraft passengers and crew need oxygen masks when an aircraft loses pressurization?

____

____

____

____

____

____

____

____

TABLE 1 *Data*

Room Temperature (K)			Atmospheric Pressure (kPa)			Syringe Diameter (cm)		
Trial	**Added Mass (kg)**	**Volume (cm^3)**	**Added Force (N)**	**Added Pressure (kPa)**	**Total Pressure (kPa)**	**Experimental *k***	**Expected *k***	**Percent Difference**
Initial	0.0000		0.00	0.00	101.3			
1	0.5000							
2	1.0000							
3	1.5000							
4	2.0000							
5	2.5000							
6	3.0000							

LAB
11C

Name

Date

Hot Space

Investigating Charles's Law

How often do you think about the pressure in the tires of your car? Most of us don't even think about the air in our tires until we discover that there is none—we have a flat tire! But this is an important maintenance issue for our cars. The pressure in our tires affects our fuel economy, which is important, but our ability to control our cars is also contingent on proper tire inflation. Improperly serviced tires can cause stopping distance to increase, decrease turning efficiency, and even result in hydroplaning on wet roads.

Many cars today have a monitoring system that indicates when tire pressure is low. Most people with this system notice that they get a low pressure warning each year as fall turns to winter—for all four tires! What is going on there? Why would the pressure on all four tires decrease in early winter? In this lab activity we will investigate the relationship between the volume of a gas and its changing temperature.

1. Describe Charles's law. Then give its equation form.

Why is my tire pressure lower on cold winter mornings?

QUESTIONS

- What happens to a gas's volume as its temperature changes?
- How can I determine the relationship between temperature and volume?
- How can I estimate the value of absolute zero?

EQUIPMENT

- hot plate
- Charles's law apparatus
- beakers, 600 mL (2)
- laboratory thermometer
- rubber bands (2)
- beaker tongs
- ice
- goggles
- hot mitts (1 pair)

2. Does it matter what units we use for temperature and volume when using Charles's law? Explain.

3. Suppose that at room temperature (25.0 °C) the volume of a syringe is 22.0 cm^3. What will the syringe volume be at 95.0 °C if the pressure is held constant?

4. What do you think would happen to the volume in Question 3 if you were to cool the sample all the way down to absolute zero? Explain.

Name ____________________

Procedure

A Using one of the beakers, start heating water to a boil. This will be used in Trial 4.

B Fill the other beaker about half full of ice and then add water until the beaker is approximately three-quarters full. This will be used in Trial 2.

C Remove the plastic tip cap from the syringe. Draw the syringe back to approximately 20 cm^3 and replace the cap. Do not remove the cap again during any part of the experiment.

D Attach the thermometer to the syringe with two rubber bands (see image at right). The bulb of the thermometer should be adjacent to the air space of the syringe. Do not obstruct the view of the volume scale graduations.

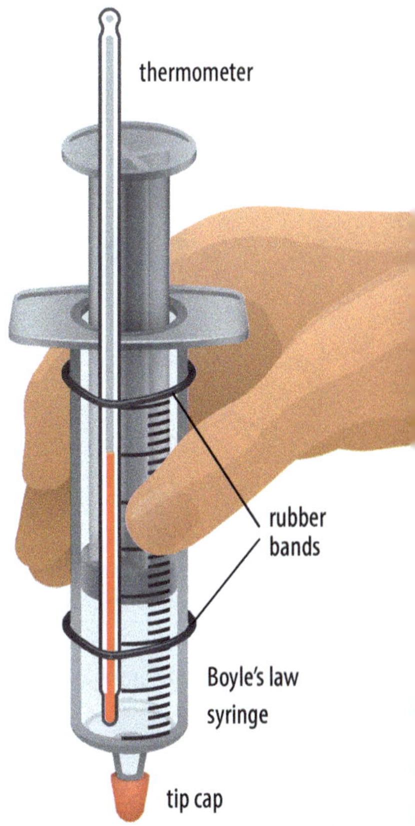

The basic Charles's law apparatus

TRIAL 1: ROOM TEMPERATURE

E Hold the syringe by the plunger shaft. Allow the syringe to sit in the air for a few minutes. Do not touch the barrel near the air space to avoid affecting the syringe air temperature or the thermometer reading with body heat.

F Measure the temperature. Convert this temperature to kelvins and record both temperatures in Table 1. This room temperature will be your T_1 when calculating the predicted volumes for Trials 2–4.

G Gently pull up on the plunger slightly and release it. Repeat this action two or three times. Note the position of the inner edge of the plunger against the scale printed on the cylinder.

H Gently push down on the plunger and release it several times. Note the position of the plunger again.

I Determine the midpoint between these two positions (Steps G and H). Record this volume as the measured volume and the predicted volume for Trial 1 in Table 1.

5. What do you think is the purpose for Steps G–I?

____

____

TRIAL 2: COLD AIR

J Check the volume of ice water in the beaker. There must be enough to cover enclosed air in the syringe barrel when it is submerged. Add cold water to adjust the level if needed.

K Place the apparatus into the ice water. (*Note:* Do not allow water to enter the plunger end of the barrel; see image below left.) Allow the temperature of the air in the syringe to reach equilibrium (3–5 minutes). Handle the apparatus only by the end of the plunger shaft and keep the air space completely submerged. Gently stirring the water will accelerate reaching thermal equilibrium.

L Once equilibrium has been reached, determine the volume, using the push-pull technique in Steps G–I. Keep the syringe submerged and avoid touching the barrel of the syringe near the air space. Record the temperatures of the water and the measured volume of the syringe in Table 1.

TRIAL 3: WARM AIR

M Empty the ice water and add hot tap water (50–70 °C) to the beaker. Allow the temperature of the air in the syringe to reach equilibrium (3–5 minutes) and then repeat Steps G–I. Record the temperatures of the water and the measured volume of the syringe in Table 1.

TRIAL 4: HOT AIR

N Using the tongs, submerge the syringe in the boiling water. The boiling water may cause the rubber bands to loosen or break, so gently hold on to both the syringe and the thermometer during this step. Allow the air temperature in the syringe to reach equilibrium (3–5 minutes) and then repeat Steps G–I. ***The syringe will be hot—use hot mitts!*** Record the temperatures of the water and the measured volume of the syringe in Table 1.

The syringe in an ice-water bath

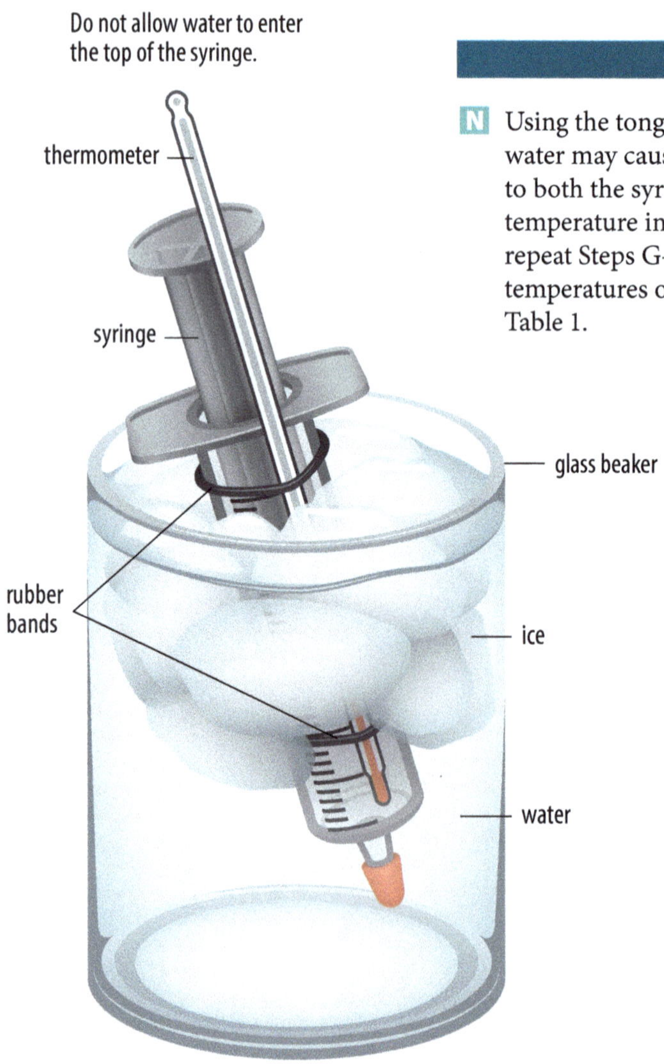

Name

Analysis

O Using the temperature and predicted volume for Trial 1, compute the predicted volume for Trials 2–4 on the basis of the temperature measured in each trial. Record these values in Table 1. Show your work for Trial 2 below.

P Calculate the percent difference between the measured and predicted volumes for each trial. Record these values in Table 1. Show your work for Trial 2 below.

Q Create a scatterplot of the measured volume versus Celsius temperature. As always, include a curve of best fit.

6. What is the relationship between volume and temperature according to your scatterplot? Does this match what you expected according to Charles's law? Explain.

____

____

____

R Use your scatterplot to extrapolate from your data to determine the x-intercept. This x-intercept is your experimental value for the Celsius temperature of absolute zero.

7. Report the temperature at the point at which the extrapolation of the graph crosses the horizontal axis.

____

8. If there were excessive friction between the plunger and the barrel, how would that affect the data?

____

____

____

____

____

Conclusion

9. Did the percent differences between the experimental and predicted values for Trials 2–4 indicate significant errors (> 5%) in the experimental data? Explain.

____

____

____

Name ______________________

10. Calculate the percent error between the extrapolated value from your data to the accepted value for absolute zero (−273.15 °C).

Going Further

11. In the opener we discussed the low pressure light coming on in cars due to decreasing temperature. Of course, this experiment was about volume and temperature. Why do you think that the car indicates a change in pressure for this situation? What does that have to do with Charles's law?

____

____

____

____

____

____

____

TABLE 1 *Charles's Law Data and Calculations*

Trial	Celsius Temperature (°C)	Kelvin Temperature (K)	Measured Volume (cm^3)	Predicted Volume (cm^3)	Percent Difference
1					n/a
2					
3					
4					

LAB
12A

Name ______________________

Date ______________________

The Capacity to Understand

Understanding Heat Capacity and Specific Heat Capacity

The fuel for many modern jet aircraft is moved through the aircraft and the engines by huge fuel pumps. Like all mechanical pumps, the fuel pumps heat up. One challenge that aircraft designers have is how to keep the fuel pumps cool. Where can they put all that excess thermal energy, especially considering all the fuel that is nearby? The answer may be surprising. Many aircraft actually use the fuel to keep the pumps cool. But fuel is flammable, so how can we use it to keep the pumps cool without starting a fire? The answer lies with the two closely related terms *heat capacity* and *specific heat capacity*. In this lab activity we will measure the heat capacity of two objects and the specific heat capacity of the material that the objects are made of.

1. Define *heat capacity*.

____

____

2. Define *specific heat capacity*.

____

____

____

What is specified in specific heat capacity?

QUESTIONS

- What is the difference between heat capacity and specific heat capacity?
- How are heat capacity and specific heat capacity related to each other?

EQUIPMENT

- hot plate
- laboratory balance
- calorimeter
- beaker, 400 mL
- density cylinders, copper and aluminum (1 ea.)
- crucible tongs
- graduated cylinder, 100 mL
- laboratory thermometers (2)
- goggles
- hot mitts (1 pair)

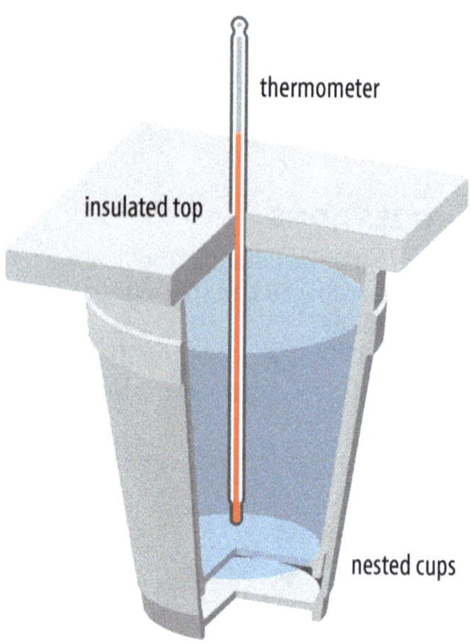

Foam-cup calorimeter

Procedure

CALORIMETER HEAT CAPACITY

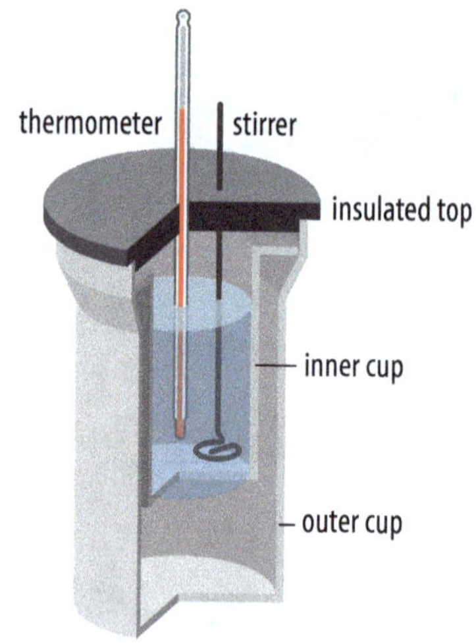

A Fill the 400 mL beaker with approximately 350 mL of tap water. Place the beaker on the hot plate and start heating the water.

B Mass each of the density cylinders and record the masses in Table 1.

C Using the tongs, place the cylinders in the beaker. ***Be careful not to drop the cylinders as they may break the beaker.*** Monitor the beaker on the hot plate. Once the water reaches boiling, note the time and then reduce the hot plate to medium-high heat to maintain a low boil.

D Remove the inner cup from the calorimeter and measure its mass. Record the mass in Table 1.

E Using the graduated cylinder, add approximately 100 mL of cold tap water to the inner cup of the calorimeter. Measure the combined mass of the calorimeter's inner cup and the water. Record this mass and the mass of the water in Table 1.

F Assemble the calorimeter and verify that the thermometer is immersed in the water but not touching the bottom of the cup.

G Measure the initial temperature of the water in the calorimeter and record your data in Trial 1 of Table 1.

Note: Perform Steps H and I quickly so that the water temperature is not significantly affected by room temperature. Keep the cover on the calorimeter except when adding a sample.

H Once the cylinders have been in the boiling water for at least five minutes, measure the hot water temperature and record this as the initial temperature of the copper in Table 1.

I Using the tongs, carefully transfer the copper cylinder from the hot water to the calorimeter. Replace the cover and check the thermometer's position in the calorimeter. Carefully stir the water constantly until the temperature stabilizes at a maximum value. Record in Table 1 the final temperature after equilibrium is reached.

3. The procedure assumes that the water temperature and the cylinder temperature are the same. Is this a good assumption? Explain.

____

____

____

J Remove the thermometer and lid from the calorimeter. Empty the inner cup of the calorimeter. Then repeat Steps E–I for the aluminum cylinder as Trial 2.

Name ____________________

Analysis

4. Which metal caused the greatest increase in the water temperature?

____

5. Use the heat formula to calculate the energy change of the water for each trial. Record your answers in Table 2. Show your work for Trial 1 below. (The specific heat capacity for water is 4.184 J/g·°C.)

6. Where did the thermal energy to heat the water come from? Explain.

____

____

____

K Record the energy lost by each metal in Table 2.

7. Calculate the heat capacity of each cylinder. Record your answers in Table 2. Show your work for Trial 1 below.

8. Which cylinder had the highest heat capacity? Make a prediction about the specific heat capacity of each metal.

____

____

____

9. Calculate the specific heat capacity of each metal. Record your answers in Table 2. Show your work for Trial 1 below.

10. Which metal had the higher specific heat capacity? Explain these results in light of your prediction.

____

____

____

11. Given that the accepted values for the specific heat capacity of copper and aluminum are 0.385 J/g·°C and 0.89 J/g·°C, respectively, calculate the percent error of each specific heat capacity that you found. Record your answers in Table 2. Show your work for Trial 1 below.

Name ____________________

Conclusion

12. What does the word *specific* refer to in *specific heat capacity*? Explain.

13. How are the concepts of specific heat capacity and heat capacity similar to density and mass?

Going Further

14. Most students will obtain values that are lower than the expected value for the specific heat capacity. What assumption was made in the procedure that would cause consistently low values? How could we modify the procedure to obtain better results?

15. The specific heat capacity of most aviation fuels is in the 2.00 J/g·°C–2.30 J/g·°C range. A jet aircraft may have its fuel pump located in a tank that contains 2.70×10^4 kg of fuel. How much energy would be needed to warm that fuel from −55.0 °C (air temperature at cruising altitude) to 38.0 °C (flash point of the fuel)? Use an average specific heat capacity value of 2.15 J/g·°C.

16. When you consider your answer to Question 15, do you think that using the fuel as a heat sink is a good design choice? Explain.

____

____

____

____

____

____

TABLE 1

Mass Inner Cup (g)				
Trial	**Item**	**Mass (g)**	**Initial Temperature (°C)**	**Final Temperature (°C)**
1	Inner Cup and Water			
	Copper			
	Water			
2	Inner Cup and Water			
	Aluminum			
	Water			

TABLE 2

Item	Trial	Energy Gained (J)
Water	1	
	2	

Item	Trial	Energy Lost (J)	Heat Capacity (J/°C)	Specific Heat Capacity (J/g·°C)	$\%_{error}$
Copper	1				
Aluminum	2				

LAB
12B

Name

Date

What's Cooking?

Building a Better Solar Oven

We all need food to survive. Some foods are good raw and may even be healthier when uncooked. Other foods need to be cooked either for safety reasons or just to make them palatable. Imagine that you are in a survival situation. How will you cook food while you wait to be rescued? Most of us would think of building a fire. What if there were no wood to burn? As you contemplate the dilemma you are in, you notice the sun beating down on you. Can you use the energy of the sun to cook food? Can you build a solar oven?

In this STEM activity you will build a solar oven to cook something assigned by the teacher. You will need to build your oven within the constraints that you are given. You will also need to develop a test for the effectiveness of your oven. Once you have tried out your oven, you will have the opportunity to modify your design to improve its efficiency.

How can I use the sun's energy to cook food?

QUESTIONS

» How does the radiation of the sun turn to thermal energy in a solar oven?

» How can I test the effectiveness of my oven's design?

» How can I improve the efficiency of my oven?

EQUIPMENT

- **as determined by your teacher**

Procedure

DESIGN PARAMETERS

Design and build a solar oven within the constraints outlined below.

- Your teacher will specify the materials that you may use. You will have access to materials for both testing and final construction.
- Your teacher will indicate the food item that you will cook in your solar oven.
- You will develop testing procedures for assessing the effectiveness of your oven.

PLANNING THE DESIGN

A Research solar ovens by doing an internet search.

B Brainstorm with your lab group on basic concepts such as how large your oven will be, how you will maximize absorption of solar energy, how you will keep the thermal energy in the oven, and what it means for an oven to be effective.

C You may need to conduct some testing to determine the best materials for your oven's basic design.

D Each member of the group should sketch his recommended design. Indicate on this plan that this is a *preliminary design*.

E Discuss with your group members the positive and negative aspects of each design and reach a consensus on the best design.

F Have a group member draw a diagram of the group's planned oven. Indicate on the diagram that this is an *initial product design*.

BUILDING THE DESIGN

G Construct your oven according to the drawing that you created in Step F.

Name

TESTING THE DESIGN

H Develop a test that will evaluate the effectiveness of the design. Some questions to consider: What are the key aspects of an effective oven, particularly a solar oven? How well should the temperature be regulated in the oven? Will an effective oven perform consistently over multiple trials?

I Conduct a test of your oven using the procedures that you developed. Use the data to demonstrate the effectiveness of your oven.

J Cook a sample of the food specified in the parameters section.

REFINING THE DESIGN

K Evaluate the performance of your oven in both the group-developed testing and the food preparation test.

L Make a drawing of your improved design. Indicate on the drawing that this is the *modified production design.*

M If time permits, build your modified oven.

RETESTING THE DESIGN

N If you built your modified oven, test it and evaluate the effectiveness of its modifications.

LAB

13

Name

Date

Microstates

Understanding Entropy

Entropy is a challenging topic. You can read of numerous definitions of the term: entropy is the measure of the disorder of a system, the measure of the dispersal of the energy of a system, or the measure of the microstates of a system. The second law of thermodynamics is a law about entropy. Since there are multiple descriptions of entropy, we have multiple statements of the second law, as we discussed in the Student Edition. In this lab activity we will look at the microstate concept of entropy. From this we will expand to some of the other descriptions of entropy.

1. Consider a single die. What are the possible microstates when rolling a single die once? What are the possible outcomes, or macrostates?

The probability of a particular outcome is the number of times that an outcome occurs divided by the total possible outcomes that can occur.

2. On the basis of your answer to Question 1, what are the probabilities of each possible outcome?

3. Using your answers from Questions 1 and 2, can you predict the outcome of the roll of a single die? Explain.

Is entropy disorder?

QUESTIONS

» How many ways can each die land when two dice are rolled?

» How many possible outcomes are there when two dice are rolled?

» How can we predict the outcome of a roll of a pair of dice?

EQUIPMENT

- dice (2), different colors

Procedure

SINGLE DIE

A Roll a single die and use a tally to record the outcome in Table 1.

B Repeat Step A until you have at least 100 rolls.

C Count up your tally marks for each microstate and record each total in Table 1.

D Create a histogram of your one-die roll outcomes.

TWO DICE

4. Consider rolling two dice. What are the possible outcomes (macrostates) consisting of the sum of the numbers shown on the dice?

____

5. What do you think is the most likely outcome? Explain.

____

____

____

____

E Roll the pair of dice and record the total count on the two dice in Table 2.

F Repeat Step E until you have at least 300 rolls.

G Create a histogram of your two-dice roll outcomes.

6. Do the histograms match your expected probabilities for the one-die and two-dice rolls? Explain.

____

____

____

____

____

Name ______________________

INVESTIGATING TWO-DICE MICROSTATES

Consider a red die and a white die. The microstates of the two-dice rolls are the roll shown on the red die and the roll shown on the white die. The macrostates are the sum of the numbers shown on the two dice. For example, you may have rolled a 1 on the red die and a 2 on the white die (the microstate), resulting in a macrostate of 3. But you could also have rolled a 2 on the red die and a 1 on the white die (microstate) for the same macrostate of 3. Let's look at what causes the shape of the two-dice histogram.

H Create a graphic organizer showing all the microstates that produce each of the macrostates for rolling the pair of dice.

7. According to your graphic organizer, how many total microstates exist for the two-dice system?

____

8. According to your graphic organizer, what is the probability of each macrostate? Show your work for the macrostate of 4.

____

____

Analysis

9. According to your answer to Question 8, what are the expected totals for each macrostate for the two-dice rolls? Record your results in Table 2. Show your work below for the macrostate of 4.

10. Calculate the percent difference for each of the outcomes in Table 2. Record your results in Table 2. Show your work below for the macrostate of 4.

11. What is the most likely outcome of a two-dice roll? Why is this true?

____

____

Name ____________________

Conclusion

12. Explain how entropy can be described on the basis of microstates.

Playing cards are often used to describe entropy in terms of the disorder. A brand new deck of cards is arranged by suit, and then each suit is in numerical order. When a deck of cards is shuffled, its entropy increases.

13. Explain this in terms of disorder. How does this relate to microstates?

Going Further

14. How would the dice-rolling situation change if we were to use three dice?

15. Consider the system of 16 gas particles shown below. If the valve were opened and the system left to itself, what would you expect to happen to the gas particles? Explain.

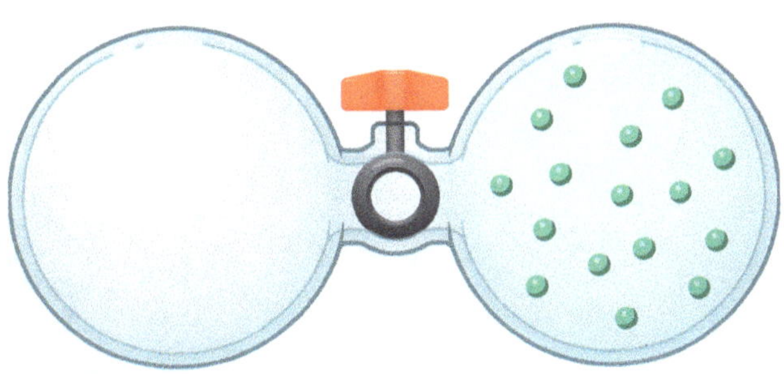

16. How likely would it be for all the particles to accumulate in the left-hand chamber? Explain.

____

____

17. If we needed to get all the particles in one chamber, what would we have to do? How does this relate to the second law of thermodynamics?

____

____

____

TABLE 1 *One-Die Rolls*

Outcome	Tally	Total
1	卌 卌 ||||	
2	卌 卌 卌 |	
3	卌 卌 卌	
4	卌 卌 卌 |||	
5	卌 卌 卌 ||	
6	卌 卌 卌 卌	

TABLE 2 *Two-Dice Rolls*

Outcome	Tally	Total	$Total_{expected}$	$\%_{difference}$
2	卌 ||			
3	卌 卌 卌 |			
4	卌 卌 卌 卌 卌 ||			
5	卌 卌 卌 卌 卌 卌 |||			
6	卌 卌 卌 卌 卌 卌 卌 卌 ||			
7	卌 卌 卌 卌 卌 卌 卌 卌 卌 卌 卌 |||			
8	卌 卌 卌 卌 卌 卌 卌 卌 |			
9	卌 卌 卌 卌 卌 卌 |			
10	卌 卌 卌 ||||			
11	卌 卌 卌 ||			
12	卌 ||||			

LAB
14A

Name

Date

Navigating Water Weighs

Understanding Buoyancy

In Genesis Chapter 6 God told Noah to build an ark. He told Noah how large to build the ark, and we would have to assume that the details He provided to Noah included how to build it so that it would float and be stable in the water.

Whether an object floats depends on its weight and how much water its volume can displace. If an object's volume displaces a volume of water that weighs less than the object's weight, the object will sink. If an object's volume displaces a volume of water with a weight exactly that of the object's weight, the object will have neutral buoyancy, which means that it will remain at whatever depth it was initially. It won't sink to the bottom and it won't float to the surface. When an object has a volume that can displace a volume of water that weighs more than it does, the object will have positive buoyancy; that is, it will float to the surface and part of the object will rise above the surface of the water. The submerged volume of the object displaces a volume of water that weighs exactly the same as the weight of the object.

Regardless of whether an object has positive, neutral, or negative buoyancy, an object submerged in water will have an apparent weight that is less than its actual weight due to the upward buoyant force produced by the water. As you can imagine, these relationships of weight and volume also imply that density is a factor of buoyancy and thus determine whether an object will float. In this lab activity we will determine the buoyant force on objects using three methods and then evaluate these methods.

How does being in a fluid affect my weight?

QUESTIONS

- How can I measure the buoyany force on a submerged object?
- How do different methods of measuring buoyant force compare?

EQUIPMENT

- triple beam balance
- support rod (laboratory bench)
- beakers, 200 mL (2)
- density cylinders (2)
- density cube
- graduated cylinder, 25 mL
- overflow can
- string
- liquid detergent

1. State Archimedes's principle in your own words.

____

____

____

Procedure

METHOD 1: DIFFERENCE IN APPARENT WEIGHT

A Place the triple beam balance on top of the support rod by inserting the rod into the hole located in the underside of the balance base (see image at right). Check that the balance is zeroed.

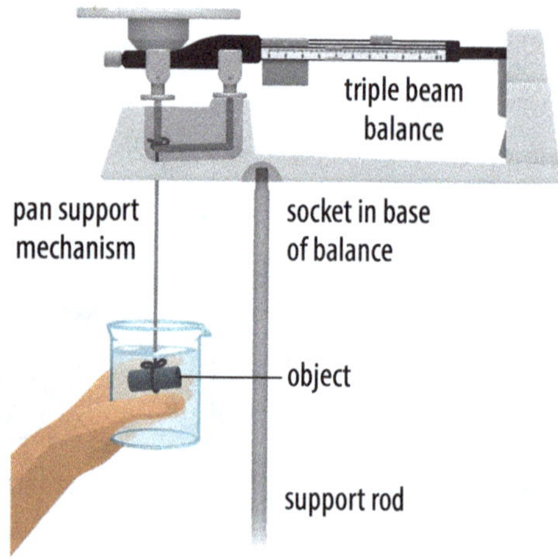

B Note the vertical support mechanism for the balance pan, which can be seen underneath the left end of the balance base. Tie a 30 cm piece of string to the lower hinge pin of the pan support mechanism.

C Fill one of the beakers with approximately 150 mL of water.

D Immerse the string in the beaker of water until it is soaked. Remove the string from the beaker and measure the mass of the wet string. Record this value in Table 1.

E Tie the string to one of the density cylinders. The cylinder should hang freely without touching anything. A slipknot will make untying easier. Record the material of the cylinder in Table 1.

F Measure the mass of the density cylinder and the string and record this value in Table 1.

2. What do you expect to happen to the balance reading when you submerge the object in water? Explain.

____

____

____

____

____

____

G Without removing the object from the balance and without touching the object in any way, completely immerse it (unless it floats) in the beaker of water. Do not allow the object to touch the container sides or bottom. If it floats, do not force it under the water. Measure the "apparent" mass of the object and string in the water and record this measurement in Table 1.

Name ______________________

3. Draw the free-body diagram for the object hanging in the water.

H Repeat Steps E–G for two other objects. Note that one or more of the objects may float. (Space is provided in Table 1 for measuring additional objects.)

I Remove the string from the balance pan support mechanism. Remove the balance from the support rod and place it on the tabletop. Verify that the balance is still zeroed.

METHOD 2: MEASURING THE DISPLACED WATER

J Measure the mass of a dry graduated cylinder. Record this mass in Table 1.

K Tie the string to the first object that you used in Method 1.

L While holding your finger over the spout of the overflow can, fill the can with water above the spout hole. Keep your finger over the spout while your partner places a drop of the liquid detergent on his fingertip, rubs the drop between two fingers, and then touches his finger to the surface of the water in the overflow can. Do not use too much detergent because bubbles will form and impede the flow of water. If you have bubbles on the surface, wait for them to disappear before proceeding.

4. What do you think is the purpose of the liquid detergent? If you need a hint, look up the word *surfactant*.

____

____

____

____

M Set the overflow can at the edge of a table and place the second beaker under the spout. Remove your finger and allow the water to flow out of the spout until it stops dripping.

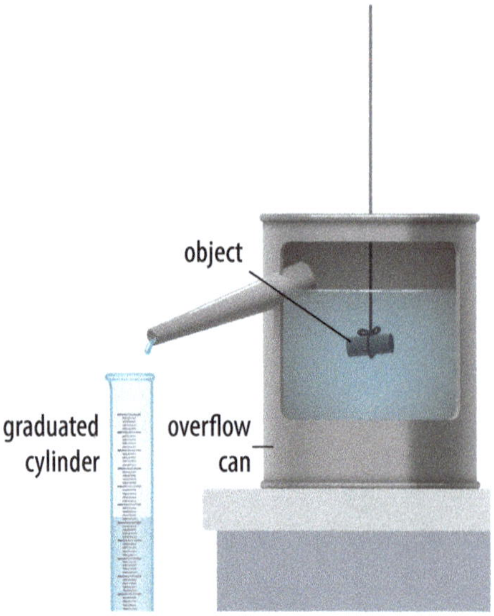

Apparatus for determining displacement volume

N After the dripping stops, remove the beaker and place the graduated cylinder under the spout. Slowly lower the object on the string into the overflow can, while catching all of the water displaced by the object (see image at left).

O Measure the mass of the graduated cylinder and water. Record the combined mass in Table 1.

P Measure the volume of the displaced water and record it in Table 1. Empty the graduated cylinder.

Q Repeat Steps K–P for the other two objects. (Once again, space is provided in Table 1 for measuring additional objects.)

R Copy the material for each object from Table 1 to Table 2.

S Calculate the weight (in air) and apparent weight (in water) of each object and record in Table 2. *Note:* You will have to factor out the mass of the string.

5. Why do you think we used the term *weight* for the object in air and *apparent weight* for when it was in water?

____

____

____

____

____

T Determine buoyant force (F_{b1}) from weight difference—the difference between the weight in air and in water. Record the result for each object in Table 2.

U Determine the mass of the water displaced by subtracting the mass of the graduated cylinder from the combined mass of the cylinder and the displaced water. Then calculate the buoyant force (F_{b2}) from the weight of the water displaced by each object, and record your result for each object in Table 2.

METHOD 3: MEASURING THE DISPLACED WATER

The buoyant force can also be calculated by using the equation $F_b = \rho g V$, where ρ is the density of the fluid (1 g/cm^3), g is the magnitude of the acceleration due to gravity, and V is the volume of the displaced water (from Table 1).

V Calculate the buoyant force (F_{b3}) from Archimedes's principle, using the above equation and record it for each object in Table 2 as F_{b3}.

Name

Analysis

W For each object find the percent difference between the two buoyant forces calculated in Steps T (F_{b1}) and U (F_{b2}). Record your results in Table 2. Show your work for Trial 1 below.

X For each object find the percent difference between the two buoyant forces calculated in Steps T (F_{b1}) and V (F_{b3}). Record your results in Table 2. Show your work for Trial 1 below.

Conclusion

6. Compare the results from the three methods used to find the buoyant force.

7. Which method do you think is most closely connected to Archimedes's principle (definition of buoyant force)?

8. A branch breaks off a waterlogged, submerged tree stump 2 m below the surface of a lake. The branch has the same density as the water around it. Will it rise or sink?

9. A raft is floating on the surface of a lake. As people start to get on the raft, it floats lower in the water. As more of the raft is pushed deeper into the water, what happens to the buoyant force acting on the object?

Going Further

10. How would being in water affect a person's weight? Explain.

11. The Civil War era USS Monitor had an empty weight of 960 metric tons.
    a. How much water did the Monitor displace?
    b. Assume that the crew and load of the Monitor weighed 43 metric tons. How much water did the Monitor displace when fully loaded?
    c. On the basis of your answer from *a* and *b*, what other changes likely occurred when the ship was loaded?

Name ______________________

12. Assume that you place a beaker of water on a scale and adjust the scale to zero. You then suspend a 85.5 g object that has a volume of 45.6 mL on a string in the water so that it is fully submerged and does not rest on the bottom of the beaker. What will the reading on the scale be? Explain.

____

____

____

____

____

____

____

____

____

TABLE 1 *Mass and Volume Data*

Mass of Wet String (g)			Mass of Graduated Cylinder (g)	
Material	**Mass of Object and String (g)**	**Mass of Object (in water) (g)**	**Mass of Cylinder and Water (g)**	**Volume Displaced Water (mL)**

TABLE 2 *Buoyant Force Calculations*

Material	Object Weight (in air) (N)	Object Apparent Weight (N)	Weight Difference (F_{b1}) (N)	Weight of Displaced Water (F_{b2}) (N)	$\%_{difference}$	Calculated Buoyant Force (F_{b3}) (N)	$\%_{difference}$

LAB
14B

Name

Date

Draft Day

Designing a Paper Boat

Lab 14A investigated the concept of buoyancy. When we think about boats, we all understand that they are supposed to be buoyant—a boat should float! Depending on the design of a ship, the percentage of ship above and below the water line varies greatly. Some ships, such as fully loaded barges, have a large percentage of their volume below the water level. In nautical terms these ships have a large *draft* (amount of ship below the water level) and a small *freeboard* (amount of the hull above the water line). Other ships, such as cruise ships, have a small draft and a large freeboard. How does a designer know whether a particular ship design will float high or low in the water?

Another aspect of shipbuilding is stability. As the percentage of ship above the water line increases, we have to be concerned with the stability of the ship. How does the designer know whether a ship will be stable? It would be embarrassing to launch a new ship and watch as it rolled over because it was top-heavy. Ship builders add weight, or ballast, to the lower hull of a ship to move the center of mass downward so that it is below the center of buoyancy. This makes for a stable design.

In this STEM activity you are going to build a boat out of paper. The challenge to this task is twofold. You must build the boat with a limited amount of supplies. You must also determine how much ballast (pennies) to launch your ship with so that it floats with 65% of the boat below the water line.

How can I design a paper boat?

QUESTIONS

» How can I design a boat with limited materials?

» How can I determine the ballast load needed to produce a particular draft?

» How can I evaluate the design of my boat?

EQUIPMENT

- computer paper (2 pcs.)
- clear plastic tape, 30 cm
- pennies

Procedure

DESIGN PARAMETERS

As you design your boat, you must work within the design parameters listed below.

- Your boat must be made from only two sheets of computer paper and 30 cm of clear plastic tape.
- You may use any boat design you want as long as its hull can be made from a single piece of computer paper. The other sheet of paper can be used to create any type of support structure for your boat.
- You will use pennies as ballast for your boat, and you must determine the number of pennies needed to cause your boat to float with 65% of the hull below the water line.

PLANNING THE DESIGN

A Keeping in mind the design parameters listed above, brainstorm possible designs for your paper boat with your team.

B Each team member should individually research how different possible designs may affect a paper boat's ability to hold pennies without sinking.

C Brainstorm with your team what information you may need about pennies. Do the necessary research and experimentation to determine needed data.

D Draw and label a diagram of your plan for your boat. Indicate that this is a *preliminary design.*

E Share your research findings with your team. Reach a consensus on the final design that your group will build and submit for testing.

F Calculate the number of pennies that will cause the boat you design to float with 65% of the hull below the water line.

G Have someone in the group draw and label a diagram of the final design. Indicate that this is an *initial production design.*

Name ______________________

BUILDING THE DESIGN

H Construct your boat according to your initial production design.

I On the side of your completed boat, mark the anticipated water line at 65%.

J Measure and record necessary data (e.g., draft, freeboard, hull height) for analysis of your boat performance.

TESTING THE DESIGN

K Load your boat with the ballast.

L Slowly lower the boat into the water so that it floats freely. Carefully lift the boat out of the water.

M The water should have left a line for the actual water line. Again measure and record necessary data (e.g., draft, freeboard, hull height) for analysis of your boat performance.

REFINING THE DESIGN

N On the basis of your testing results, make modifications to your boat and ballast load, if needed. If changes are made, draw and label a new diagram indicating that this is the *final production design*. Include a description of how any proposed changes to the original design are intended to improve the performance of the new design.

RETESTING THE DESIGN

O If time permits, build and test your modified design.

P On the due date submit your preliminary, initial production, and final production designs as well as your analysis regarding the performance of your boat.

LAB

15A

Name

Date

Are You Positive?

Inquiring into Electric Charge

Copier machines work by generating a static electric force that binds particles of powdered toner to paper. This is the same force that causes the static cling that makes socks stick to sweatshirts in a clothes dryer. Static electricity causes the surprising shock generated after walking across a carpeted floor on a dry winter morning, and that same static buildup generates lightning bolts that can cause great damage during summer thunderstorms. There are two types of charge, positive and negative. Most of the time there is a balance between these two types of charge, but when there is an imbalance of charge on an object, we say that the object has a static charge. In this inquiry activity you will design an experiment to determine the type of charge on an object.

How can I determine the type of charge on an object?

QUESTIONS

- What experimental procedures can test for the type of charge on an object?
- How can I put a charge of known type on an object?
- How should objects of like and unlike charges interact?

Procedure

PLANNING/WRITING SCIENTIFIC QUESTIONS

A Research static electricity, static charge, and the triboelectric effect.

B Write a hypothesis about how a charged object will react when it is brought near a positive object or a negative object. Write a second hypothesis about how you will be able to identify an object as having either a positive or negative charge.

C Write a series of scientific questions that will guide your experimental design to test your hypotheses from Step B.

DESIGNING SCIENTIFIC INVESTIGATIONS

D Work with your lab group to determine how you can answer your scientific questions and test your hypotheses.

E Write your procedures and have your teacher approve them.

CONDUCTING SCIENTIFIC INVESTIGATIONS

F Conduct your investigation according to your procedures.

DEVELOPING MODELS

G Develop a model for how objects of like and different charges react to each other. The model should include how you can know the charge on an object that you electrify. The model should also include how objects of unknown charge will interact with objects of known charge.

SCIENTIFIC ARGUMENTATION

H State claims about your hypotheses from Step B. As always, support your claims with specific evidence from your data or analysis.

I How could you improve the design of your experiment to produce more accurate results?

LAB

15B

Name ______

Date ______

Simply Shocking

Investigating Electric Charge

Most of us are familiar with static cling, but what is going on with the girl shown above? Of course, static cling occurs because materials can gain excess positive or negative charge, and we know that opposites attract. So how does static electricity explain hair going in all directions as it is doing in the image? In this case each strand of hair has gained the same type of excess charge. Because each strand of hair has the same charge, each strand is repelling every other strand!

How do objects gain excess charge? There are three primary means of charging an object: friction, conduction, and induction. In this lab activity we will investigate the interactions between charged objects as well as the different methods of charging.

1. State the law of charges.

______

______

2. Will rubbing an object with a piece of wool be most likely to leave the object with a positive or a negative charge?

______

______

______

______

How do charged objects interact?

QUESTIONS

» How can I charge objects by induction and conduction?

» How can I determine whether a material is a conductor or an insulator?

EQUIPMENT

- electroscopes (2)
- friction rod kit
- rod stirrup
- support stand
- lightweight string or thread
- masking tape
- conductor

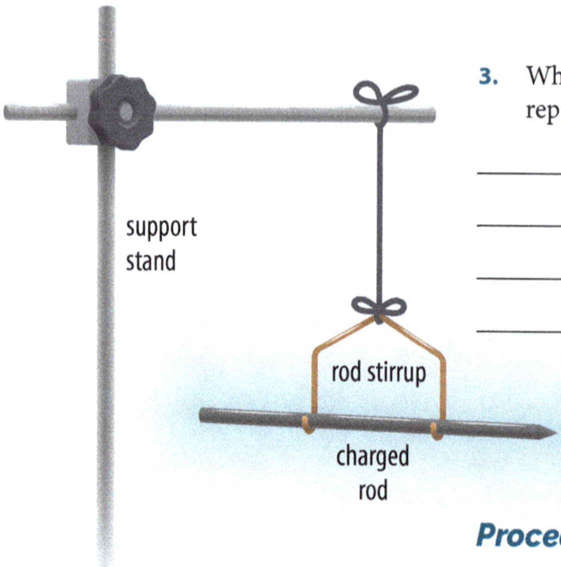

3. When dealing with static electricity, what can we conclude if two objects repel each other? if they attract each other?

____

____

____

____

Most Positive *(readily lose electrons)*

- rabbit's fur
- acrylic
- acetate
- glass
- quartz
- mica
- human hair
- nylon
- rayon
- wool
- cat's fur
- silk
- paper
- cotton
- wood
- sealing wax
- amber
- rubber
- metals
- polyester
- polystyrene (foam)
- plastic wrap
- polyurethane
- polyethylene
- polypropylene
- sulfur
- celluloid
- vinyl (PVC)
- teflon

Most Negative *(readily steal electrons)*

Procedure

PART 1: DETECTING CHARGES

A Suspend the rod stirrup from the support stand with a string.

B Wrap a piece of masking tape around one end of the rubber rod. Hold the taped end of the rod and charge it by rubbing with wool. Place the charged rod into the stirrup (see above). A rubber rod rubbed with wool should be negatively charged. This is your known rod. The other items to be tested—rubber, acrylic, and glass—will be called test objects.

C Charge a second rubber rod—your first test object—with wool.

D Bring the test rod near the known rod, but do not allow them to touch. If the rods do touch, you will need to recharge the known rod. If the rubber rod does not repel the known rod, recharge the known rod before proceeding.

E Charge each of the three test objects first with silk and then with wool as indicated in Table 1. On the basis of how the test object interacts with the known rod, record the charge on each test object in Table 1. If the rods touch, you will need to recharge the known rod before proceeding.

F Remove the rubber rod from the stirrup. Touch the stirrup in order to remove any excess charge that may have built up. You may also want to touch your hand to a ground point, such as a water faucet, in order to remove any excess charge buildup on your body. Even the wool and silk should be grounded by touching them to a ground point.

G Now charge the acrylic rod with silk and suspend the acrylic rod in the stirrup. To confirm the charge on the acrylic rod, rub another acrylic rod with silk and notice whether it is attracted or repelled. An acrylic rod rubbed with silk should be positively charged.

H Charge each of the three test objects first with silk and then with wool as indicated in Table 1. On the basis of how the test object interacts with the known rod, record the charge on each test object in Table 1. Remember that if the rods touch, you will need to recharge the known rod before proceeding.

Name ____________________

PART 2: CHARGING BY INDUCTION AND CONDUCTION

I Ground one of the electroscopes. Record the charge on the electroscope and the behavior of the leaves in Table 2.

4. How do you know the charge on the electroscope?

____

____

____

____

J Rub the rubber rod with wool. Bring the rod near the electroscope sphere without touching it. Record the charge on the electroscope and the reaction of the foil leaves in Table 2.

K Now touch the rod to the electroscope sphere and remove it. Bring the rod near the electroscope sphere again. Record the reaction of the foil leaves in Table 2.

5. Why did the leaves behave as they did in Steps J and K?

____

____

____

____

____

L Ground the electroscope. Record the charge on the electroscope and behavior of the leaves in Table 2.

M Rub the acrylic rod with silk. Bring the rod near the electroscope sphere without touching it. Record the reaction of the foil leaves in Table 2.

N Now touch the rod to the electroscope sphere and remove it. Bring the rod near the electroscope sphere again. Record the reaction of the foil leaves in Table 2.

6. Did the leaves behave differently in Steps M and N compared with J and K? Explain.

____

____

____

____

O Charge a rubber rod with wool and bring it close to the charged electroscope. Record the reaction of the foil leaves in Table 2.

7. Explain what happened to the electrons in Step O.

____

____

____

PART 3: TESTING CONDUCTORS AND INSULATORS

8. If you were to connect the spheres of two electroscopes with a conductor and then were to charge one of the electroscopes, (a) how would you expect the leaves of both to behave? (b) What if you did the same experiment, but connected the spheres with an insulator?

Testing conductors and insulators

P Set the two electroscopes side by side. Ground each of the electroscopes and then suspend a conductor from the rod stirrup so that it touches both electroscope spheres. Charge a rubber rod with wool and touch one of the spheres. Record the reaction of the leaves of both electroscopes in Table 3.

Q Ground each of the electroscopes, and then repeat Step P using a nonconductor (rubber, acrylic, or glass rod) to connect the two electroscope spheres. Record the reaction of the leaves of both electroscopes in Table 3.

Analysis

9. The end of the rod that you held in your hand in Step B is likely to be neutral. Why do you think this is true?

10. If the end of the rod that you held in Step B is neutral, how can there be an excess charge on the other end of the rod?

11. Explain how wool produces a negative charge on rubber but produces a positive charge on acrylic.

12. Did the leaves behave as expected in Part 3?

Name ______________________

Conclusion

13. Describe how charging by friction, conduction, and induction occur.

14. Explain which parts of the procedure involved charging by friction, conduction, and induction.

Going Further

15. Which forms of charging do you think are involved in the production of lightning?

TABLE 1 *Part 1 Data*

Known Negative Object

Known rod is	rubber	Charge is	negative

Test Object	Silk		Wool	
	Attract or Repel?	Charge	Attract or Repel?	Charge
Rubber				
Acrylic				
Glass				

Known Positive Object

Known rod is	acrylic	Charge is	positive

Test Object	Silk		Wool	
	Attract or Repel?	Charge	Attract or Repel?	Charge
Acrylic				
Rubber				
Glass				

TABLE 2 *Part 2 Data*

Step	Charge on Rod	Charge on Electroscope	Behavior of Leaves
I	n/a		
J	negative		
K	negative		
L	n/a		
M	positive		
N	positive		
O	negative		

TABLE 3 *Part 3 Data*

Material	Behavior of Leaves
Conductor	
Insulator	

LAB
16A

Name

Date

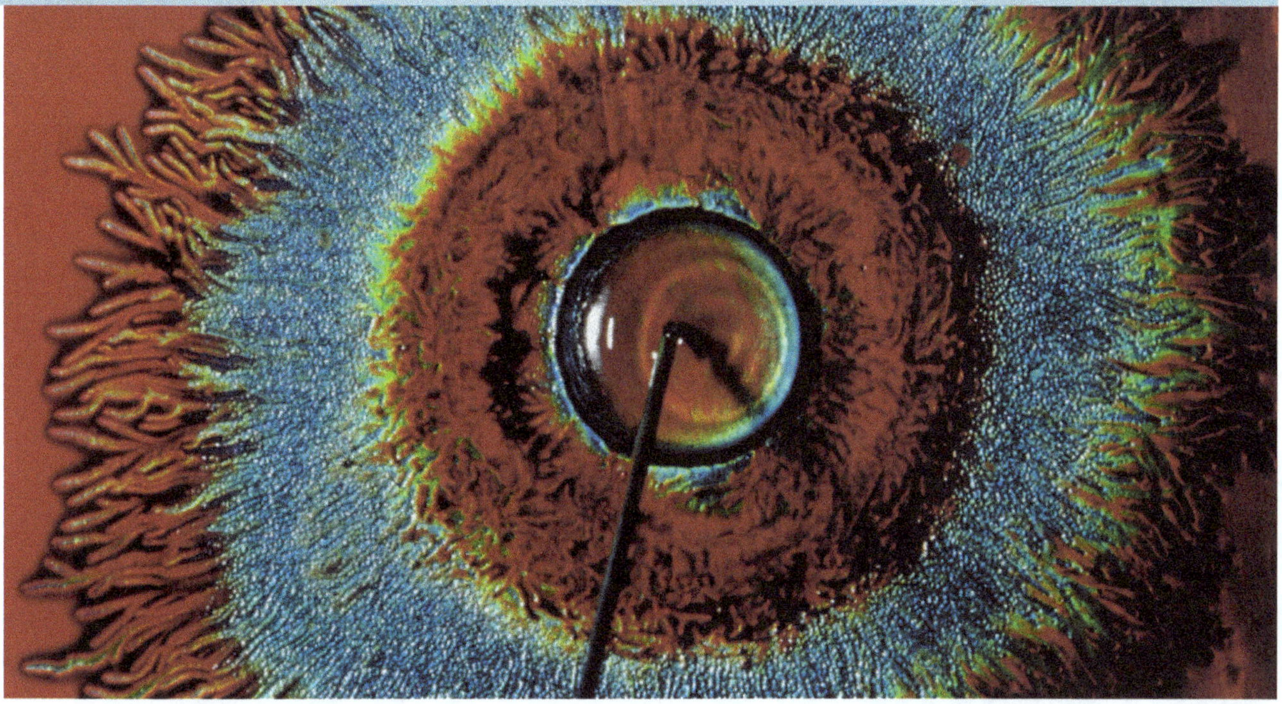

Field Work

Mapping Electric Fields

A field force is one that works at a distance, meaning that objects can interact without being in contact with each other. We all are familiar with gravitational force. And while we can't see gravity, it is easy to visualize the field because it is pointing in toward the center of a massive object. In our everyday experience the gravitational field is always pointing toward the center of the earth. The earth holds us firmly on the ground, or in some cases unceremoniously flings us to the ground. As we saw in Chapters 15 and 16, electric fields are similar to gravitational fields but are more challenging because electric fields point toward positively charged objects and away from negatively charged objects. In this lab activity we will use the concept of equipotential surfaces to map the field between pairs of charged objects.

1. Describe an electric field.

2. Explain why a test charge is a small positive charge.

How can I visualize the field between two charged objects?

QUESTIONS

- How can I identify an equipotential surface with a voltmeter?
- How can I map an electric field if I know the equipotential surfaces around the object?
- What determines the shape of the electric field between charged objects?

EQUIPMENT

- DC power supply
- multimeter
- conductive ink pen
- conductive paper
- insulating platform
- pushpins (2)
- lead wires (4)
- colored pencils (3), blue, green, and red
- masking tape

3. Define *equipotential surface.*

____

____

Procedure

PART I: DETECTING CHARGES

A Use the conductive ink pen to draw the conductors on conductive paper as shown in Drawing Area 1.

B In Drawing Area 1 use the blue colored pencils to sketch your prediction of the electric field lines between the conductors.

C Attach the conductive paper to the insulating platform with masking tape.

D Insert one pushpin into each of the conductors drawn on the conductive paper. Leave some of the metal post of the pushpin exposed.

E Attach two lead wires to the power supply by plugging one wire into the negative terminal and the other wire into the positive terminal. Connect each of the free ends of these lead wires to the posts of the pushpins.

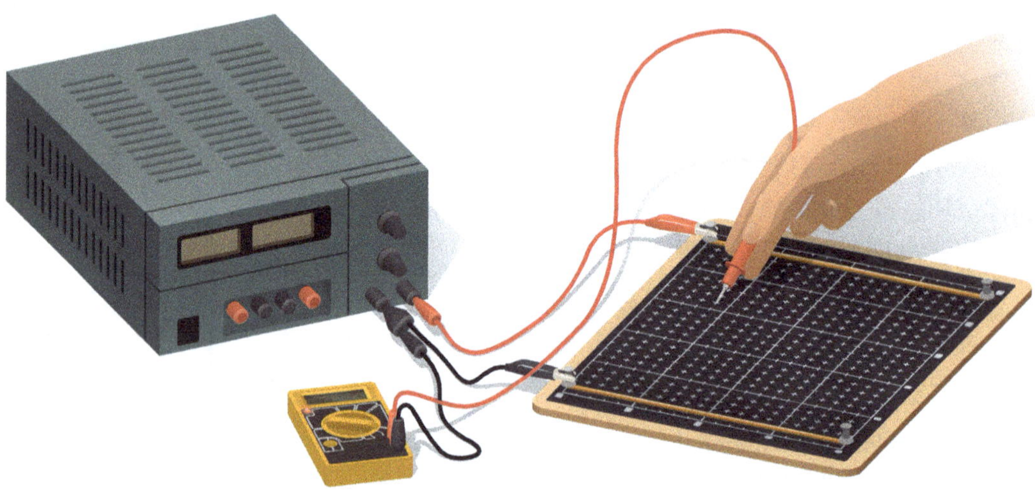

F Using a third lead wire, connect the com port of the multimeter to the negative terminal of the power supply. Connect the fourth lead wire into the VΩmA port of the multimeter. Your setup should look like the image above.

G Turn the power supply to Volts DC and adjust the voltage to 14 V.

H Select DCV 20 V on the multimeter.

I Using the free end of the lead wire from the VΩmA port of the multimeter, locate a position near the left side of the conductive paper with a voltage of 2.0 V. Using the red pencil, plot this position in Drawing Area 1.

J Find four more locations with a voltage of 2.0 V and plot the positions in Drawing Area 1. Try to cover the width of the paper.

K Repeat Steps I and J for voltages of 4.0 V, 6.0 V, 8.0 V, 10.0 V, and 12.0 V.

L Repeat Steps A–K for the conductor setup shown in Drawing Area 2.

Name ______________________

Analysis

M In each of the Drawing Areas use the green pencil to connect the points of equal potential—the equipotential surfaces.

N In each of the Drawing Areas use the red pencil to draw in the actual field lines. Field lines and equipotential surfaces are always perpendicular to each other.

4. Why do you think field lines and equipotential surfaces meet at right angles?

____

____

____

5. Compare your predicted electric field and the actual electric field. How well were you able to predict the shape of the electric field?

____

____

Conclusion

6. Where do you think the field is strongest and weakest within each field? Explain.

____

____

____

Going Further

7. What would be the electric potential energy of an object of −5.4 μC on the 6.0 V equipotential surface of Drawing Area 1?

8. How much work would you have to do to an object of −5.4 μC to move it from one position with voltage 6.0 V to a different position with a voltage of 6.0 V on Drawing Area 1? Explain.

9. How much work would you have to do to an object of −5.4 μC to move it from a position with voltage 6.0 V to a different position with a voltage of 2.0 V on Drawing Area 1? Explain.

10. What would be the magnitude of the electric field strength on an object of −5.4 μC on the 6.0 V equipotential surface of Drawing Area 1? You will need the distance from the 0 V potential.

Name ____________________

11. What would be the magnitude of the electric force on an object of $-5.4\ \mu C$ on the 6.0 V equipotential surface of Drawing Area 1?

12. In what direction would the electric force be on an object of $-5.4\ \mu C$ on the 6.0 V of Drawing Area 1?

____

____

Drawing Area 1

Name ______________________

Drawing Area 2

LAB

16B

Name ______________________

Date ______________________

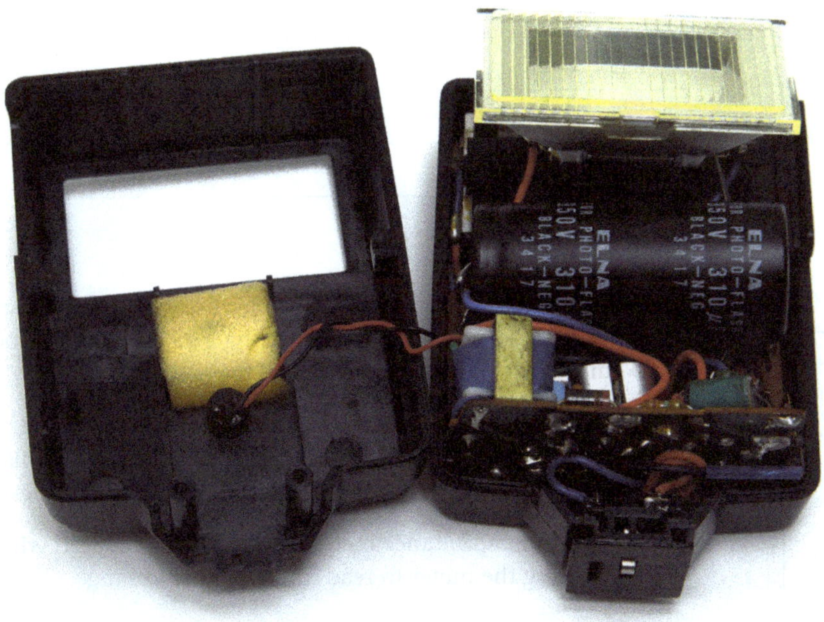

Charge It

Investigating Capacitance

A camera flash is simply a light bulb, but of course it turns on for a very short amount of time and then turns off. How is this done? Where does the electrical energy for such a short, powerful burst come from? We are all familiar with batteries that power so many of our modern electronic devices. Batteries actually store chemical energy, which is converted to electrical energy when needed. Batteries work well to light a small bulb, but the bulb usually stays lit for a while. To get a short burst of light, camera flashes typically use capacitors. A camera contains a charging circuit that prepares the capacitor. A separate discharge circuit uses the stored electrical energy. The amount of charge that a capacitor can store depends on the geometry and materials used to build the capacitor. In this lab activity we will investigate the effect of varying the geometry of a parallel plate capacitor.

1. What is a capacitor?

____

____

2. What is the relationship between capacitance, voltage, and charge?

____

____

____

How can I store charge to be used later?

QUESTIONS

» How is the capacitance of a parallel plate capacitor related to its geometry?

» How can I evaluate my measured capacitance?

EQUIPMENT

- capacitance meter
- scissors
- metric ruler
- aluminum foil 20 cm × 28 cm (2 pcs.)
- lead wires (2)
- physics textbook

3. Describe a parallel plate capacitor.

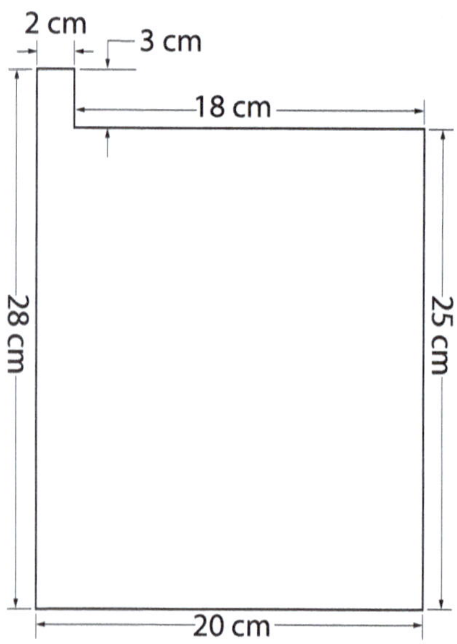

Procedure

A Make the capacitor plates by cutting a 3 cm × 18 cm piece off the short side of each piece of foil, leaving a 25 cm × 20 cm foil plate with a 3 cm × 2 cm tab as shown at left.

B Measure the thickness of the stack of pages 101 through 220 (60 sheets) in your textbook when the pages are pressed together. Calculate the average thickness per sheet in meters. Record this average thickness in Table 1.

C Set up your capacitance meter by connecting the lead wires to the meter and setting the meter to read in the picofarad range.

CHANGING PLATE SEPARATION

In this part of the procedure we will investigate the effect that changing plate separation has on the capacitance. The area of the two plates will remain constant but the plates will be moved closer together.

4. What effect do you think decreasing the separation between the plates will have on the capacitance? Explain.

D Place one of your aluminum plates between pages 100 and 101 with the tab close to the spine and sticking out the top of the book.

E Place the second plate between pages 220 and 221 with the tab close to the outer edge of the book and sticking out the top of the book. This will separate your plates by 60 sheets of paper. Make sure that you are aligning the two plates when placing them between the pages. Close the book.

Name ______________________

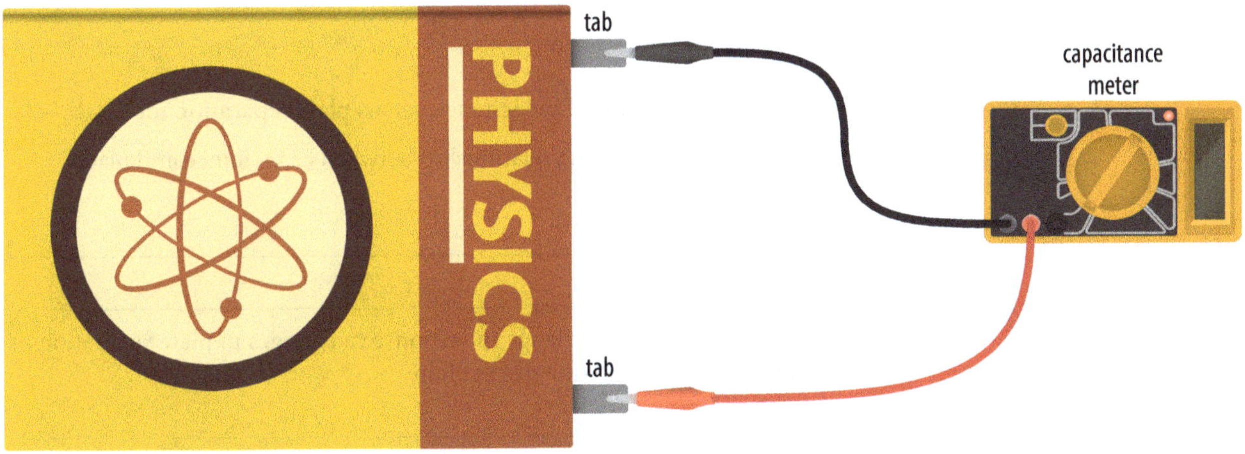

F Connect the two leads from the meter to the two tabs of the capacitor. Press down on the cover of the book and measure the capacitance. Record the capacitance in Table 1.

G Repeat Steps E and F with the second plate between pages 200 and 201 (50-page separation), pages 180 and 181 (40-page separation), pages 160 and 161 (30-page separation), pages 140 and 141 (20-page separation), and pages 120 and 121 (10-page separation).

CHANGING PLATE AREA

In this part of the procedure we will investigate the effect that changing plate area has on the capacitance. The separation between the two plates will remain constant. Make sure that you are aligning the two plates when placing them between the pages. Plate 1 will remain between pages 100 and 101 and Plate 2 will remain between pages 120 and 121.

5. What effect do you think decreasing the area of the plates will have on the capacitance? Explain.

____

____

____

H Disconnect the two leads from the capacitor. Remove Plate 1 and fold the plate in half so that its dimensions match those given in Trial 7 of Table 1. Return Plate 1 to its position between pages 100 and 101 with the tab near the spine and sticking out the top of the book.

I Remove Plate 2 and fold the plate in half so that its dimensions match those given in Trial 7 of Table 1. Return Plate 2 to its position between pages 120 and 121 with the tab toward the outer edge of the book and sticking out the top of the book. Make sure that you are aligning the two plates when placing them between the pages. Close the book.

J Connect the two leads from the meter to the two tabs of the capacitor. Press down on the cover of the book and measure the capacitance. Record the capacitance in Table 1.

K Repeat Steps H–J, folding the plates to match the dimensions given for Trials 8–11. Make sure that you are aligning the two plates when placing them between the pages.

Analysis

L Create a scatterplot of capacitance versus plate separation for Trials 1–6.

6. State a claim about the relationship between capacitance and plate separation.

____

____

7. Did the capacitance change in response to changes in plate separation as you expected (Question 4)? Explain.

____

____

____

M Create a scatterplot of capacitance versus plate area for Trials 6–11.

8. State a claim about the relationship between capacitance and plate area.

____

____

9. Did the capacitance change in response to plate area as you expected (Question 5)? Explain.

____

____

____

N Copy the measured capacitance for Trial 6 from Table 1 to Table 2.

O Calculate the plate separation in meters for Trial 6. Show your work below and record in Table 2.

Name

P Given that the dielectric constant for this textbook paper is 3.7 and the permittivity of free space is 8.854 pF/m, calculate the expected capacitance for Trial 6 and record in Table 2.

Q Calculate the percent error for Trial 6 and record in Table 2.

Conclusion

10. Describe how the geometry of a parallel plate capacitor affects its capacitance.

Going Further

11. A camera flash must store 0.141 C of charge at 300.0 V. The camera uses a parallel plate capacitor with 8.21 m^2 plates separated by a porcelain dielectric ($\kappa = 6.5$). What is the separation between the plates?

Name

TABLE 1 *Data*

<table>
<tr><th colspan="2">Thickness per Sheet (m)</th><td colspan="4"></td></tr>
<tr><th>Trial</th><th>Separation (pages)</th><th>Length (m)</th><th>Width (m)</th><th>Area (m^2)</th><th>Capacitance (pF)</th></tr>
<tr><td>1</td><td>60</td><td rowspan="6">0.2500</td><td rowspan="6">0.2000</td><td rowspan="6">0.050 00</td><td></td></tr>
<tr><td>2</td><td>50</td><td></td></tr>
<tr><td>3</td><td>40</td><td></td></tr>
<tr><td>4</td><td>30</td><td></td></tr>
<tr><td>5</td><td>20</td><td></td></tr>
<tr><td>6</td><td rowspan="6">10</td><td></td></tr>
<tr><td>7</td><td>0.2500</td><td>0.1000</td><td>0.025 00</td><td></td></tr>
<tr><td>8</td><td>0.1250</td><td>0.1000</td><td>0.012 50</td><td></td></tr>
<tr><td>9</td><td>0.1250</td><td>0.0500</td><td>0.006 25</td><td></td></tr>
<tr><td>10</td><td>0.0625</td><td>0.0500</td><td>0.003 125</td><td></td></tr>
<tr><td>11</td><td>0.031 25</td><td>0.0500</td><td>0.001 563</td><td></td></tr>
</table>

TABLE 2 *Evaluation*

Trial	Measured Capacitance (pF)	Plate Separation (m)	Plate Area (m)	Expected Capacitance (pF)	$\%_{error}$
6			0.0400		

LAB
17A

Name ______________________

Date ______________________

Circuit Court

Understanding Circuits and Resistors

We are surrounded by electrical devices, and in most cases they work well and make our lives better. Electrical devices are built around electric circuits. The purpose of a circuit is to move electrical energy from the power source to the components of the circuit. When devices work, it is because the circuit directs the proper current and energy to where it belongs and keeps it from where it does not belong. When circuits fail, the results can range from disappointing to disastrous. In the late 1980s a military refueling aircraft caught fire during landing because of a short circuit—an unintended path—in its electrical system. After the insulation around a wire was damaged, the electric current moved out of the wire and into a fuel tank, igniting the fuel vapors. The aircraft was destroyed, but thankfully all the crew members survived.

This lab activity will provide you a good introduction to circuits and the working of resistors in circuits. We will build basic circuits with resistors. We will compare the expected and measured values of the resistors that we use. We will then investigate the relationships between the current, voltage, and resistance through a resistor in a circuit. This lab activity should spark your interest in circuits!

How are current, voltage, and resistance related?

QUESTIONS

» How can I use a circuit diagram to build a circuit?

» How accurate are the labels on resistors?

» What happens to current when resistance increases?

» What happens to current when voltage increases?

EQUIPMENT

- **multimeters (2)**
- **DC power supply**
- **resistors (5)**
- **lead wires (4)**

1. Draw a simple circuit connecting a 15.0 V battery, a switch, a lamp, and a resistor in a single path.

2. What is electrical resistance? What causes resistance?

3. What is a resistor?

4. In what direction does current flow in a circuit?

Procedure

INVESTIGATING RESISTORS

A Observe the color bands on each of the resistors and record the colors in order in Table 1. The first color band is always closest to the end of the resistor. With small resistors it is often difficult to see that one end is farther from the color band than the other end. Gold or silver cannot be the first band, so if you find a gold band on one end, start from the other end of the resistor.

Name ______________________

B Refer to the resistor color key at right to identify the resistance of each of your resistors. The first two or three bands represent the significant digits of the resistance. The third or fourth band represents the power of 10 by which the significant digits are multiplied (the "multiplier"). Record this as the expected resistance for each resistor in Table 1. For additional help decoding resistor color codes, see Appendix I.

C Decode the tolerance (fourth or fifth band), or the permissible variation of the actual resistance compared to its expected values, expressed as a percentage. Record the tolerance in Table 1.

D Set up one of the multimeters to read resistance. You are using it as an ohmmeter at this point. Select the indicator range according to the expected resistance that you decoded in Step B.

E Measure the resistance of each of your resistors by simultaneously touching the multimeter leads to the opposite ends of the resistors. Record the measured resistance in Table 1.

Resistor Color Codes

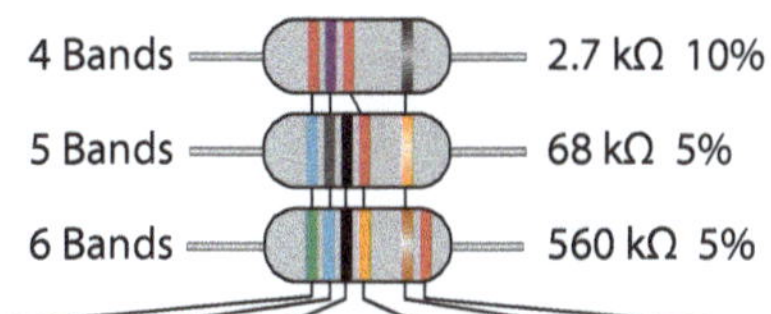

1st Digit	2nd Digit	3rd Digit	Multiplier	Tolerance	Temperature Coefficient
0	0	0	1		
1	1	1	10	1%	
2	2	2	100	2%	
3	3	3	1 k		
4	4	4	10 k		
5	5	5	100 k	0.5%	
6	6	6	1 M	0.25%	
7	7	7	10 M	0.1%	
8	8	8	0.01	10%	
9	9	9	0.1	5%	

INVESTIGATING CURRENT AND RESISTANCE

F Set up one multimeter as a voltmeter, selecting a range that can measure approximately 10 V. Set up the other multimeter as an ammeter, selecting a range that can measure approximately 100 mA.

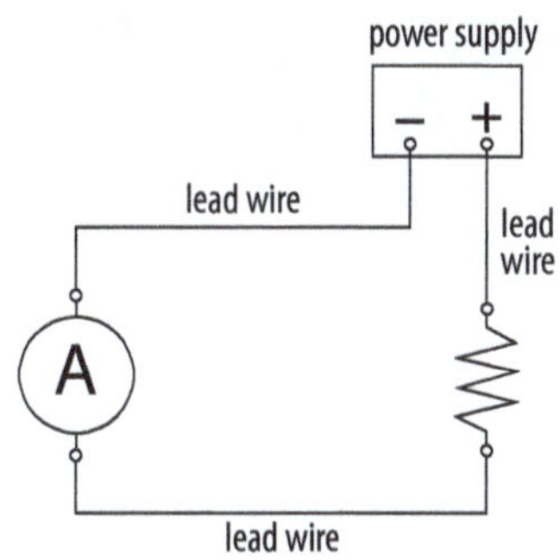

G Build the circuit as shown in the diagram at right. Check that the power supply voltage selector is set to 0. Record the expected resistance (as decoded in Step B) of the resistor that you are using in Table 2.

H Turn on the power supply and set the voltage to approximately 6.0 V. Use the voltmeter to measure the voltage across the two terminals of the power supply and record this supplied voltage value in Table 2. ***Resistors can overheat. If the resistor begins to smoke, turn off the power supply and notify your teacher.***

I Measure the voltage drop across the resistor by touching the two leads from the voltmeter, one on each side of the resistor (see image at right). Record this value in Table 2.

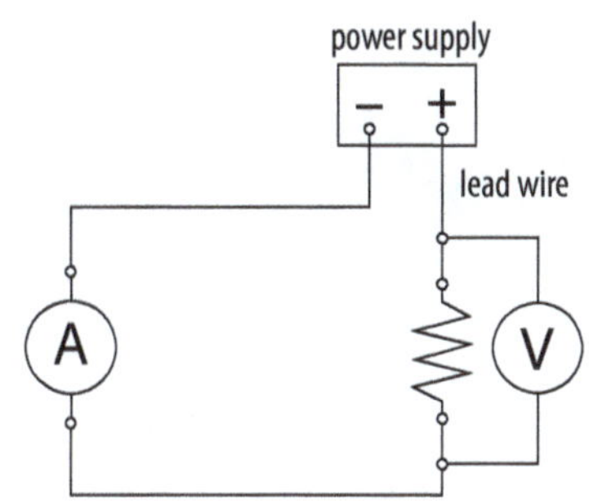

J Read the current on the ammeter and record in Table 2.

K Turn off the power supply.

L Replace the resistor with each of the other four resistors. Turn on the power supply and repeat Steps I–K.

INVESTIGATING CURRENT AND VOLTAGE

M Replace the resistor in the circuit with the fourth-highest-valued resistor. Record the indicated value of the resistor that you are using in Table 3.

N Turn on the power supply.

O Set the voltage to 15.0 V. Use the voltmeter to measure the voltage drop across the resistor by touching the two leads from the voltmeter, one on each side of the resistor. Record this value in Table 3.

P Read the current on the ammeter and record in Table 3.

Q Repeat Steps O and P using voltages of 12.5 V, 10.0 V, 7.50 V, 5.00 V, 2.50 V.

R Turn off the power supply.

Analysis

INVESTIGATING RESISTORS

5. Consult the user manual for the multimeter. What is the tolerance for the ohmmeter, voltmeter, and ammeter?

6. Calculate the percent error between the measured resistance and the expected resistance. Record your answers in Table 1. Show your work for the first resistor below.

INVESTIGATING OHM'S LAW

7. How did the voltage drop across the resistor compare with the voltage supplied in each trial? Explain.

S Create a scatterplot of current versus resistance from Table 2.

8. Using your scatterplot from Step S, describe the relationship between current and resistance. Include the equation for the best-fit curve and the R^2 value.

Name

T Create a scatterplot of current versus voltage from Table 3.

9. Using your scatterplot from Step T, describe the relationship between current and voltage. Include the equation for the best-fit curve and the R^2 value.

10. Use Ohm's law to calculate the resistances for each trial in Tables 2 and 3. Enter your results in Tables 2 and 3. Below, show your work for the first resistor in Table 2.

Conclusion

11. Are your answers to Questions 8 and 9 consistent with Ohm's law?

12. How accurate were the resistance values marked on the resistors compared to the measured values of resistance?

Going Further

13. In the event of a short circuit, how do you think the current compares to the current in the circuit as designed?

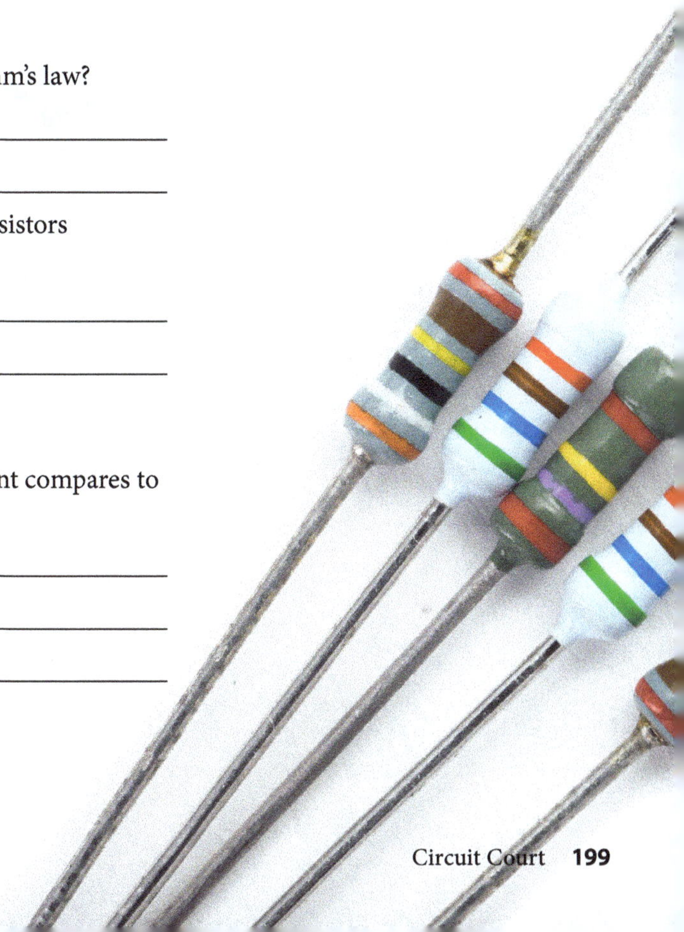

TABLE 1 *Resistance Values*

Color Code	Expected Resistance (Ω)	Tolerance (%)	Measured Resistance (Ω)	Percent Error

TABLE 2 *Changing Resistance*

Supplied Voltage (V)			
Expected Resistance (Ω)	**Voltage Drop (V)**	**Current (mA)**	**Resistance (Ohm's Law) (Ω)**

TABLE 3 *Changing Voltage*

Expected Resistance (Ω)	Voltage Drop (V)	Current (mA)	Resistance (Ohm's Law) (Ω)

LAB

17B

Name ____________________

Date ____________________

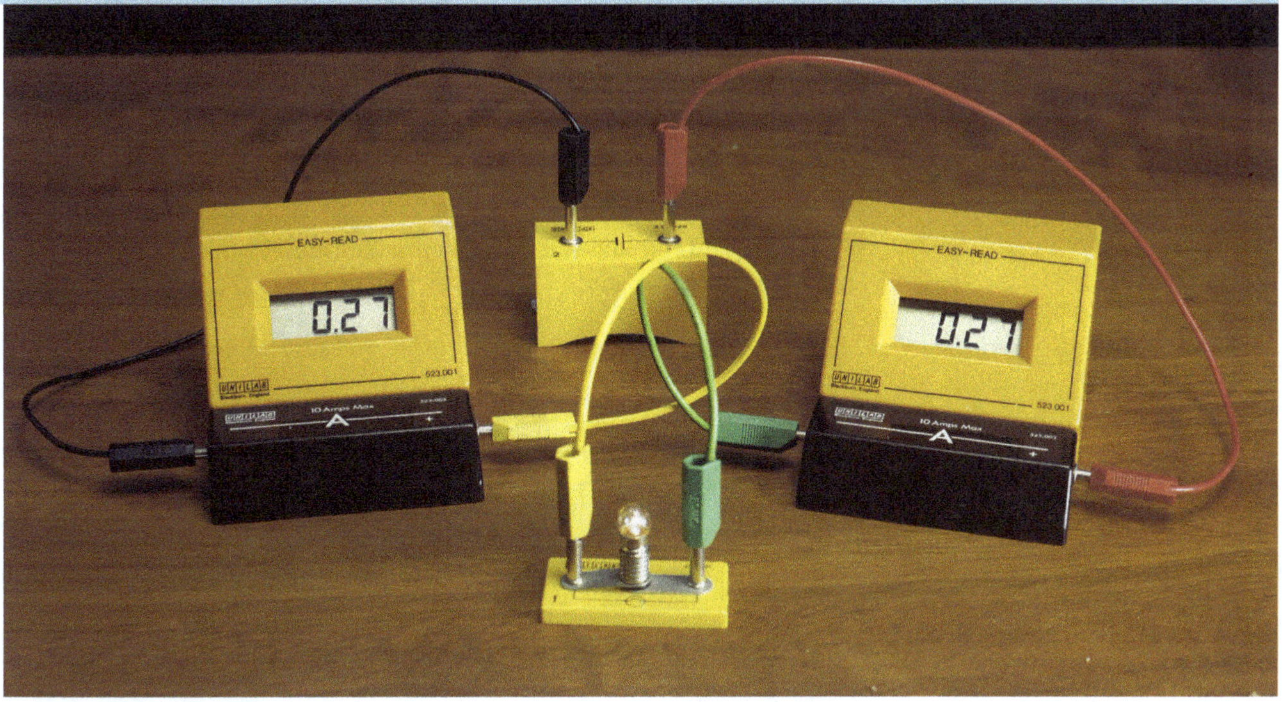

The Lineup

Investigating Series Circuits

Living in the digital age, we are accustomed to the fact that many devices we use on a daily basis operate on electricity. Most of these devices are simple to operate. We make sure that they are connected to a power supply, we adjust a few settings, and the device operates flawlessly. But what is going on behind the scenes, or should we say, *within* the scenes? What happens inside a device that makes it operate properly?

The internal workings of our electronic devices are composed of electric circuits. Circuits move electricity from a power supply, through a variety of paths, to and through different components within a device. Each of those components has a specific range of operating values for voltage and current. The circuit is designed to adjust the current and voltage that is available to each of the components.

Simple circuits may be classified as *series* or *parallel*. Series circuits are circuits in which current has only one path to follow. Parallel circuits allow current to branch so that there is more than one path to follow. In this lab activity we will investigate the effect of connecting resistors in a single line—a series circuit. We will see how this affects the voltage drop across and current through each resistor.

1. Compare a series-connected circuit to a parallel-connected circuit.

____

____

____

____

How does connecting resistors in series affect the operation of a circuit?

QUESTIONS

- » How does connecting resistors in series affect the voltages in a circuit?
- » How does the current change as resistors are connected in series?
- » How is equivalent resistance related to individual resistor resistances?

EQUIPMENT

- multimeters (2)
- DC power supply
- resistors (4)
- lead wires (6)

Procedure

Use the circuit diagram at left to answer Questions 2–3.

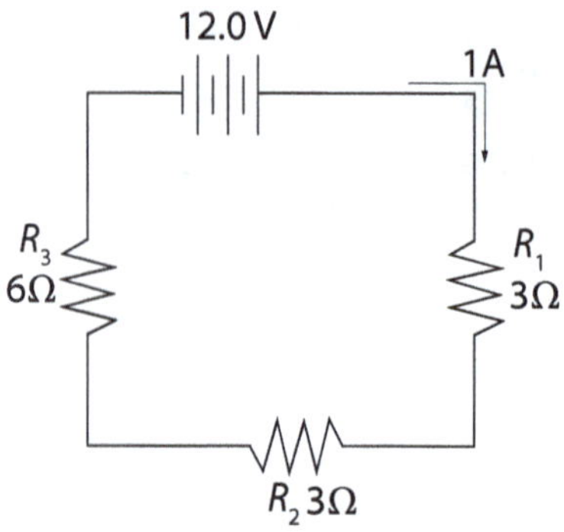

2. How much current would you expect in Resistor 1? in Resistor 2? in Resistor 3? Explain.

3. What voltage drop would you expect across Resistor 1? across Resistor 2? across Resistor 3? Explain.

CHANGES TO CURRENT AND VOLTAGE

A Refer to the resistor color key in Appendix I. Identify the resistance of each of your resistors. The first two or three bands represent the significant digits of the resistance. The third or fourth band represents the power of 10 by which the significant digits are multiplied (the "multiplier"). Record this as the expected resistance for each resistor in Table 1. You have three identical resistors and then one that is different from the others. The three identical resistors will be Resistors 1–3, and the unique one is Resistor 4.

B Set up one multimeter as a voltmeter, selecting a range that can measure at least 10 V. Set up the other multimeter as an ammeter, selecting a range that can measure approximately 100 mA.

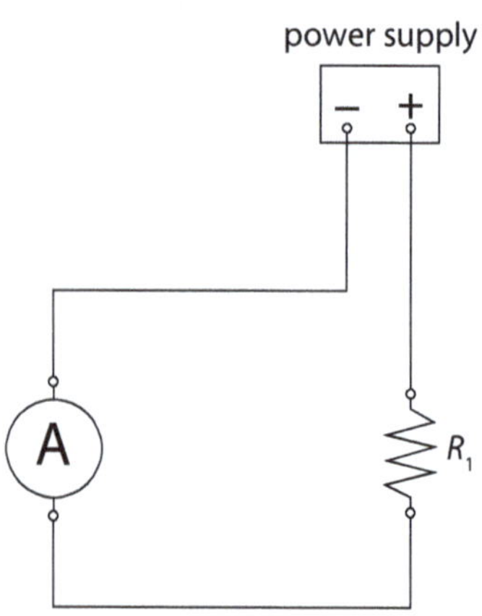

C Build the circuit as shown in the diagram at left. Check that the voltage selector is set to 0.

D Turn on the power supply and set the voltage to approximately 10.0 V. Measure the voltage across the two terminals of the power supply and record this supplied voltage value in Table 1. ***Resistors can overheat. If the resistor begins to smoke, turn off the power supply and notify your teacher.***

E Measure the voltage drop across the resistor by touching the two leads from the voltmeter, one on each side of the resistor. Record this value in Table 1.

Name ______________________

F Read the current value on the ammeter and record in Table 1.

G Have a lab partner watch the reading on the ammeter as you connect and disconnect the voltmeter across the resistor.

H Turn off the power supply.

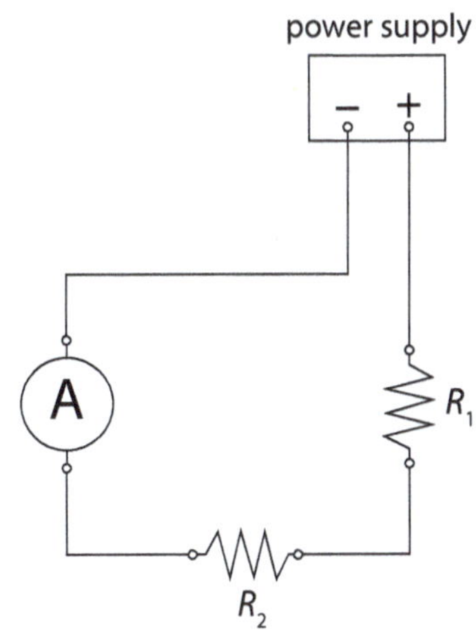

4. Does connecting and disconnecting the voltmeter affect the current within the circuit? If so, by how much? Explain.

____

____

____

____

I Add another resistor between the first resistor and the ammeter.

J Turn on the power supply and check that the voltage is set to approximately 10.0 V. Measure the voltage across the two terminals of the power supply and record this supplied voltage value in Table 1.

K Use the voltmeter to measure the voltage drop across each resistor and record these values in Table 1.

L Read the current value on the ammeter and record in Table 1.

M Turn off the power supply.

N Repeat Steps J–M two more times, each time adding a resistor between the existing resistors and the ammeter until you have a total of four resistors in the circuit. Your circuit should look like the one at right. Note that the ammeter is in Position A.

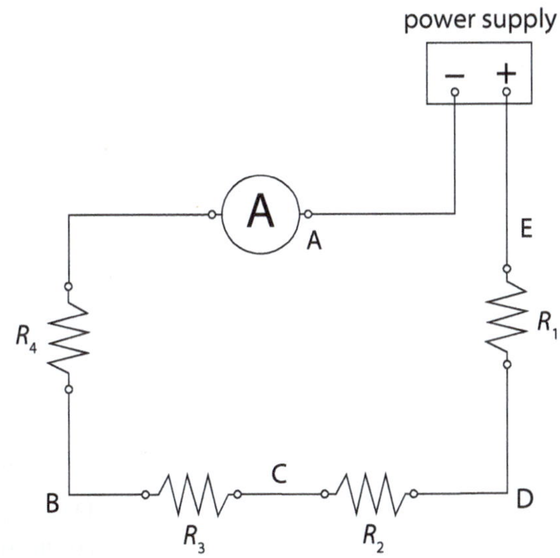

5. Do you expect the current to be different anywhere else in the circuit? Explain.

____

____

____

____

O Sequentially reconfigure the circuit so that you can measure the current at positions B, C, D, and E. Remember to turn off the power supply when changing the circuit.

6. Did the circuit act as you had anticipated in Question 5? Explain.

____

____

____

EQUIVALENT RESISTANCE

P Using Ohm's law with the supplied voltage and measured current, calculate the equivalent resistance for each set of resistors. Record your results in Table 1. Show your work for the four-resistor circuit below.

Q Reconfigure the voltmeter as an ohmmeter.

R Disconnect the four resistors in series from the circuit, but leave them connected to each other in series.

S Measure the equivalent resistance in the four-resistor series. Record the value in Table 1.

T Remove Resistor 4. Measure and record the equivalent resistance in the three-resistor series.

U Repeat Step S for the two- and one-resistor series.

Name ______________________

Analysis

7. How does the voltage drop across Resistor 1 in the one-resistor circuit compare to the supplied voltage? Explain.

8. How does the voltage drop across Resistors 1 and 2 in the two-resistor circuit compare to each other and to the supplied voltage? Explain.

9. How does the voltage drop across Resistors 1, 2, and 3 in the three-resistor circuit compare to each other and to the supplied voltage? Explain.

10. How does the voltage drop across Resistors 1 and 4 in the four-resistor circuit compare to each other? Explain.

11. How does the sum of the voltage drops across Resistors 1–4 in the four-resistor circuit compare to the supplied voltage? Explain.

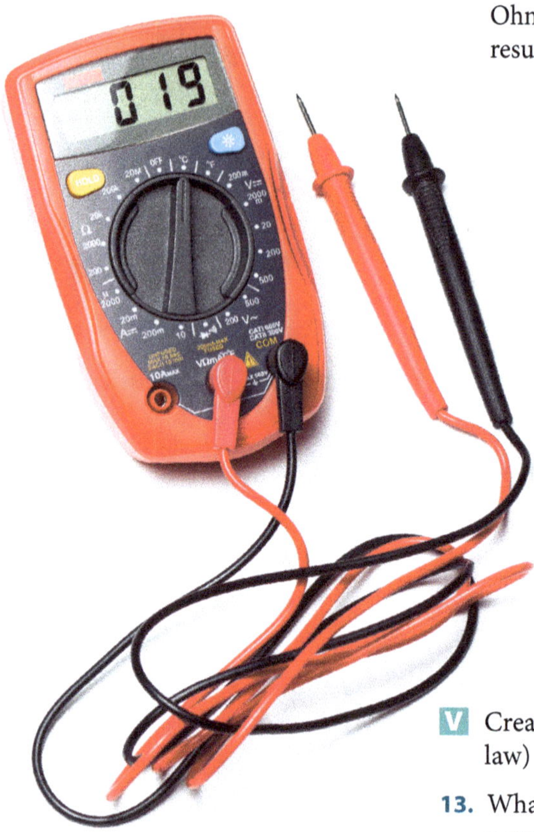

12. Calculate the percent difference between the resistance calculated using Ohm's law with the resistance measured with the meter. Record your results in Table 1. Show your work for the four-resistor circuit below.

V Create a scatterplot of current and equivalent resistance (from Ohm's law) for the four circuits.

13. What is the relationship between current and equivalent resistance in a series circuit? Include the equation for the best-fit curve and the R^2 value.

____

____

____

Conclusion

14. How does the equivalent resistance of components in series relate to the individual resistances? Write a general mathematical formula for this relationship.

____

____

15. How does current change throughout a series circuit? Write a general mathematical formula for this relationship.

____

____

16. How does voltage drop across an entire circuit relate to the voltage drop across each component? Write a general mathematical formula for this relationship.

____

____

Name ________________

17. Think back to how we measured voltage in the circuit. Was the voltmeter connected in series with the resistors? If not, how was it connected? Why do you think we connected it like this?

____

____

____

18. Think back to how we measured current in the circuit. Was the ammeter connected in series with the resistors? If not, how was it connected? Why do you think we connected it like this?

____

____

____

Going Further

19. Assume that you are required to measure the current through a resistor in an electrocardiogram machine. You can see the resistor on the circuit board and can make any necessary measurements at its terminals, but you cannot disconnect the resistor or turn off the circuit. Describe how you could make the determination under these circumstances by using only a multimeter.

____

____

____

____

TABLE 1 *Changing Resistance*

One-Resistor Circuit

Supplied Voltage (V)	

Resistor	Expected Resistance (Ω)	Voltage Drop (V)	Current (mA)	Resistance (Ohm's Law) (Ω)	Resistance (meter) (Ω)	Percent Difference
1						

Two-Resistor Circuit

Supplied Voltage (V)	

Resistor	Expected Resistance (Ω)	Voltage Drop (V)	Current (mA)	Resistance (Ohm's Law) (Ω)	Resistance (meter) (Ω)	Percent Difference
1						
2						

Three-Resistor Circuit

Supplied Voltage (V)	

Resistor	Expected Resistance (Ω)	Voltage Drop (V)	Current (mA)	Resistance (Ohm's Law) (Ω)	Resistance (meter) (Ω)	Percent Difference
1						
2						
3						

Four-Resistor Circuit

Supplied Voltage (V)	

Resistor	Expected Resistance (Ω)	Voltage Drop (V)	Current (mA)	Resistance (Ohm's Law) (Ω)	Resistance (meter) (Ω)	Percent Difference
1						
2						
3						
4						

LAB
17C

Name ______________________

Date ______________________

The Path Less Traveled

Investigating Parallel Circuits

Think about your home. Most homes have multiple appliances that operate with electricity. Have you ever thought about how the outlets in your house are wired? Are they connected in series or in parallel? How would they operate differently if they were connected in series or in parallel?

Recall that a series circuit is one that has one path for current, while a parallel circuit has multiple paths for current. For charges to move through a circuit, the circuit must be closed, usually with a switch. We turn on a coffee maker by pushing a switch. Will it work even if the microwave is turned off? Lab 17B allowed you to investigate series circuits, and this lab activity will allow you to investigate parallel circuits.

1. Think about the discussion in the opener. Do you think the coffee maker and microwave are connected in series or in parallel? Explain.

____

____

____

How does connecting resistors in parallel affect the operation of a circuit?

QUESTIONS

- How does connecting resistors in parallel affect the voltages in a circuit?
- How does the current change as resistors are connected in parallel?
- How is equivalent resistance related to individual resistor resistances?

EQUIPMENT

- multimeters (2)
- DC power supply
- resistors (4)
- lead wires (4)

Procedure

Use the circuit diagram at left to answer Questions 2–3.

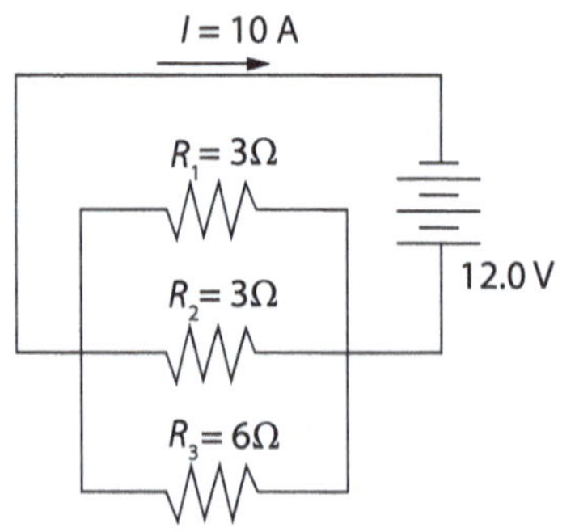

2. How much current would you expect in Resistor 1? in Resistor 2? in Resistor 3? Explain.

3. What voltage drop would you expect across Resistor 1? across Resistor 2? across Resistor 3? Explain.

CHANGES TO CURRENT AND VOLTAGE

A Refer to the resistor color key in Appendix I. Identify the resistance of each of your resistors. The first two or three bands represent the significant digits of the resistance. The third or fourth band represents the power of 10 by which the significant digits are multiplied (the "multiplier"). Record this as the expected resistance for each resistor in Table 1. You have three identical resistors and then one that is different from the other. The three identical resistors will be Resistors 1–3, and the unique one is Resistor 4.

B Set up one multimeter as a voltmeter, selecting a range that can measure at least 10 V. Set up the other multimeter as an ammeter, selecting a range that can measure approximately 100 mA.

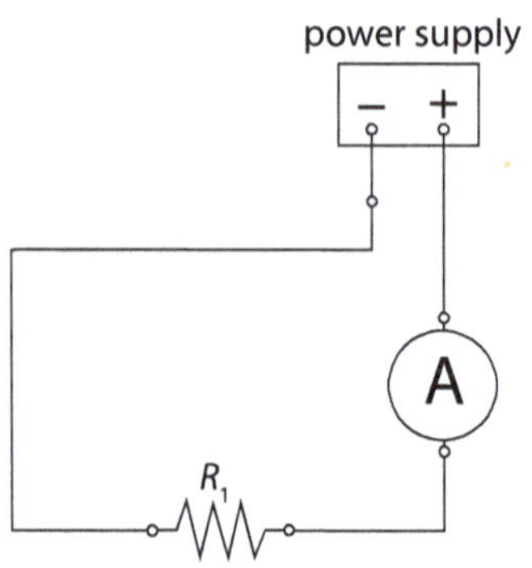

C Build the circuit as shown in the diagram at left. Check that the voltage selector is set to 0.

D Turn on the power supply and set the voltage to approximately 10.0 V. Measure the voltage across the two terminals of the power supply and record this supplied voltage value in Table 1. ***Resistors can overheat. If the resistor begins to smoke, turn off the power supply and notify your teacher.***

E Measure the voltage drop across the resistor by touching the two leads from the voltmeter, one on each side of the resistor. Record this value in Table 1.

Name ____________________

F Read the current and record in Table 1.

G Have a lab partner watch the reading on the voltmeter as you connect and disconnect the voltmeter across the resistor.

H Turn off the power supply.

4. Does connecting and disconnecting the voltmeter affect the current within the circuit? If so, by how much? Explain.

____

____

____

____

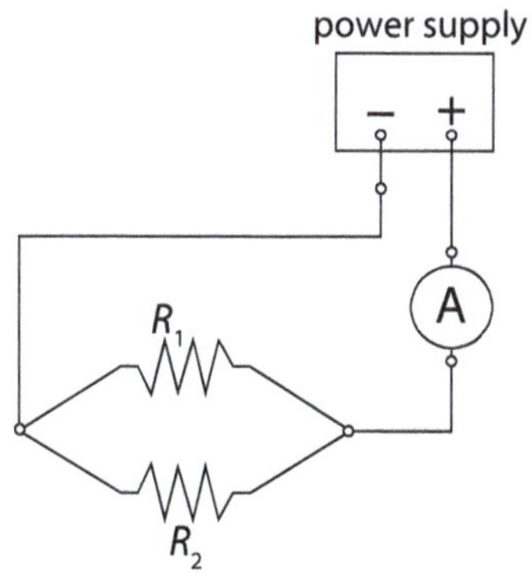

I Add another resistor next to the first resistor in between the ammeter and power supply. The circuit should match the diagram shown at right.

J Turn on the power supply and check that the voltage is set to approximately 10.0 V. Measure the voltage across the two terminals of the power supply and record this supplied voltage value in Table 1.

K Measure the voltage drop across each resistor and record these values in Table 1.

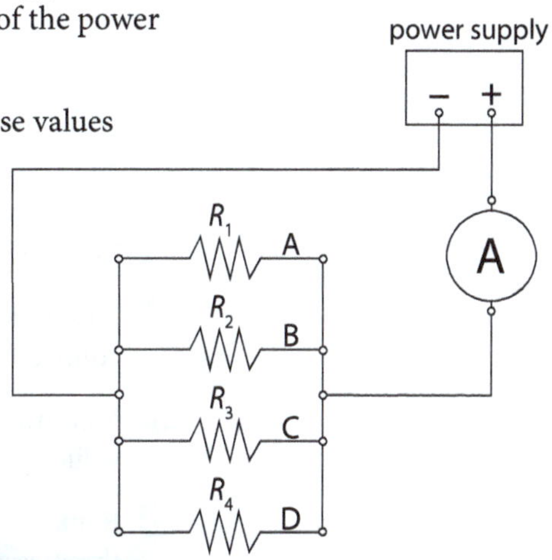

L Read the current and record in Table 1. Adjust the range on the ammeter as needed.

M Turn off the power supply.

N Repeat Steps J–M two more times, each time adding a resistor next to the existing resistors until you have a total of four resistors in the circuit. Your circuit should look like the one at right.

CURRENT IN INDIVIDUAL PARALLEL RESISTORS

O Copy the voltage drop readings from the four-resistor portion of Table 1 to Table 2.

P Sequentially reconfigure the circuit with the ammeter located at Positions A, B, C, and D so that you can measure the current through each of the four individual resistors. Remember to turn off the power supply when changing the circuit. Record your amperages in Table 2.

EQUIVALENT RESISTANCE

Q Using Ohm's law with the supplied voltage and measured current, calculate the equivalent resistance for each set of resistors. Record your results in Table 1. Show your work for the four-resistor circuit below.

R Reconfigure the voltmeter as an ohmmeter.

S Disconnect the four resistors in parallel from the circuit but leave them connected to each other in parallel.

T Measure the equivalent resistance in the four-resistor series. Record the value in Table 1.

U Remove Resistor 4. Measure and record the equivalent resistance in the three-resistor group.

V Repeat Step S for the two- and one-resistor groups.

Analysis

5. How does the voltage drop across the resistors compare to the supplied voltage? Explain.

____

____

6. What happened to the total current in the circuit as resistors were added in parallel? Explain.

____

____

____

Name ____________________

7. How do the currents in Resistors 1–3 in the four-resistor circuit compare to each other? Explain.

____

____

____

8. What is the relationship between the currents in Resistors 1–3 and the total current through the three-resistor circuit (Table 1)? Explain.

____

____

____

9. How do the currents in Resistors 1 and 4 in the four-resistor circuit compare to each other? Explain.

____

____

____

10. How does the sum of the currents in Resistors 1–4 in the four-resistor circuit compare to the total current? Explain.

____

____

____

11. Calculate the percent difference between the resistance calculated using Ohm's law with the resistance measured with the meter. Record your results in Table 1. Show your work for the four-resistor circuit below.

Conclusion

12. As resistors are added in parallel, what happens to the equivalent resistance? to the total current? to the voltage?

13. How does the equivalent resistance of components in parallel relate to the individual resistances? Write a general mathematical formula for this relationship.

14. How does total current in a parallel circuit change throughout as resistors are added? Write a general mathematical formula for this relationship.

15. How does voltage drop across an entire circuit relate to the voltage drop across each component? Write a general mathematical formula for this relationship.

16. Since we connect voltmeters in parallel, what can we conclude about the internal resistance in a voltmeter? Explain.

17. Since we connect ammeters in series, what can we conclude about the internal resistance in an ammeter? Explain.

Name ______________________

Going Further

18. As more parallel components are connected to a circuit, the total current in the circuit increases. What is the risk when there is too much current in a circuit?

____

____

19. What do electricians use to reduce the risk of fire in electric circuits?

____

TABLE 1 *Total Circuit*

One-Resistor Circuit

Supplied Voltage (V)	

Resistor	Expected Resistance (Ω)	Voltage Drop (V)	Current (mA)	Resistance (Ohm's Law) (Ω)	Resistance (meter) (Ω)	Percent Difference
1						

Two-Resistor Circuit

Supplied Voltage (V)	

Resistor	Expected Resistance (Ω)	Voltage Drop (V)	Current (mA)	Resistance (Ohm's Law) (Ω)	Resistance (meter) (Ω)	Percent Difference
1						
2						

Three-Resistor Circuit

Supplied Voltage (V)	

Resistor	Expected Resistance (Ω)	Voltage Drop (V)	Current (mA)	Resistance (Ohm's Law) (Ω)	Resistance (meter) (Ω)	Percent Difference
1						
2						
3						

Four-Resistor Circuit

Supplied Voltage (V)	

Resistor	Expected Resistance (Ω)	Voltage Drop (V)	Current (mA)	Resistance (Ohm's Law) (Ω)	Resistance (meter) (Ω)	Percent Difference
1						
2						
3						
4						

TABLE 2 *Individual Components*

Four-Resistor Circuit

Resistor	Expected Resistance (Ω)	Voltage Drop (V)	Current (mA)
1			
2			
3			
4			

LAB
18

Name

Date

Navigating Magnetism

Mapping a Magnetic Field

A field is a three-dimensional region of space where some measurable quantity, usually a force, energy, or a distribution of particles, varies from point to point within the region. A magnetic field is the region of magnetic forces surrounding a source of magnetism. Earth is surrounded by a magnetic field that is produced deep within the planet, most likely from its molten iron core. God designed the magnetic field to protect life on Earth from the radiation and high-energy particles within the solar wind.

Over time people discovered the magnetic field. We quickly learned that in addition to protecting us, the magnetic field also provides a method for navigating over great distances. While God gave man the ability to use Earth's magnetic field, He also enabled many animals, including humpback whales, turtles, birds, and bees, to use the same magnetic field. In this lab activity we will model magnetic fields using magnetic field lines. A similar technique was used in Lab 16A, in which we mapped electric fields.

In what direction does Earth's magnetic field point?

QUESTIONS

» How can I determine the direction of a magnetic field?
» How can I use field lines to model the field around a magnet?
» In what direction does the magnetic field around a current-carrying wire point?

EQUIPMENT

- Ampère's rule apparatus
- DC power supply
- bar magnets (2)
- books (2)
- cardboard sheet
- metric ruler
- small magnetic compass
- wires with alligator clips (2)
- masking tape
- computer paper (3 pcs.)

1. Describe the orientation of a compass needle in a magnetic field.

___

___

2. Discuss at least three factors that could affect the mapping of a bar magnet's field in this lab exercise.

___

___

___

3. Discuss at least one method that you could use to demonstrate that a magnetic field around a magnet is three-dimensional.

___

___

___

4. One way to view magnetic field strength is to note how close together the magnetic field lines are. Where do you expect the field lines to be closest together?

___

___

Procedure

THE FIELD AROUND ONE MAGNET

A Place one of the bar magnets on the tabletop. Set two books of equal thickness along both sides of the magnet. Place the cardboard over the books so that it is centered over the magnet. Fasten the cardboard to the books with tape so that it cannot move (see image below).

B Draw a line along all four edges of a sheet of paper, parallel to and about 2 cm from the edges.

C Mark dots at the four corners where the lines intersect. Mark five evenly spaced dots on the shorter lines at each end of the paper and eight evenly spaced dots on the longer lines at each side. Tape the sheet of paper to the cardboard.

D Position the compass so that one of the ends of the compass needle is directly over one of the dots.

E Mark another dot on the paper on the opposite side of the compass case in line with the needle. Remove the compass. Draw a line connecting the two dots. Place an arrowhead on the line pointing in the direction that the north pole of the compass was pointing.

F Position the compass so that the needle is aligned with the end of the line that you drew in Step E. Mark the paper at the opposite end of the needle (as you did above) and draw another line and arrowhead.

G Continue making line segments in the same way until the line either runs off the paper or the compass needle becomes jammed because it is pointing downward toward the magnet. If there is room, start at the original dot and extend the line in the other direction. Your paper should look something like the image shown right, top. When you have connected a dot at the edge of the paper to the magnet with a continuous line, repeat the process for each dot around the perimeter of the paper.

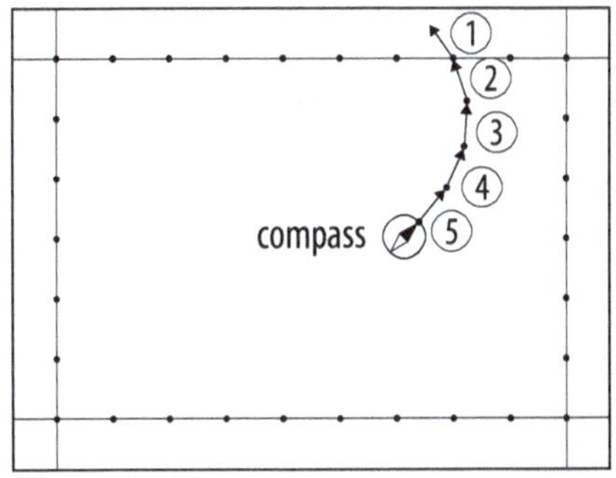

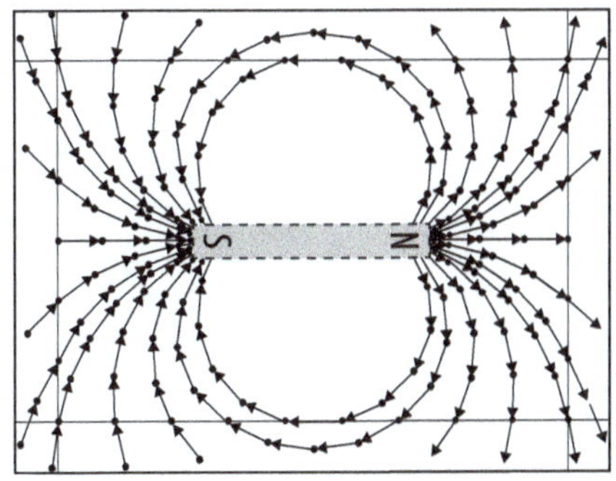

H Outline the location of the magnet with a dotted line and label the north and south poles. The completed project should look like the image shown right, bottom.

OPPOSITE-POLE INTERACTION

I Remove the paper from the cardboard and the cardboard from the books. Position two bar magnets so that the north pole of one magnet is facing the south pole of the other magnet about 15 cm away centered in the space between the two books.

J Replace the cardboard and tape a new sheet of paper to the cardboard.

K Draw a single line across the paper so that it is between the two magnets and perpendicular to their lengths.

L Mark dots on the line at 2 cm intervals.

M Starting at the dots, construct the magnetic field lines in both directions as outlined in Steps D–G.

LIKE-POLE INTERACTION

N Remove the paper from the cardboard and the cardboard from the books. Reverse one of the magnets so that the north pole of one magnet is about 15 cm from the north pole of the other.

O Replace the cardboard and tape a new sheet of paper to the cardboard.

P Draw two lines on the paper. Both lines will be parallel to the magnets about 2 cm from the long edges of the paper.

Q Mark dots on the lines at 2 cm intervals.

R Starting at the dots, construct the magnetic field lines in both directions as outlined in Steps D–G.

OUTSIDE FACTORS INFLUENCING A MAGNETIC FIELD

S Remove the cardboard and books and place one of the magnets on the table. Remove all other magnetic material to a remote location in the room.

T Hold the compass in your hand near the south pole of the magnet. Rotate the case until the north mark on the compass rose is aligned under the north end of the needle. Gradually move directly away from the magnet along a line from north to south along the magnet, keeping the north mark pointed at the magnet. In the blanks provided below, record the number of degrees that the compass needle moves away from the north mark at the specified distances.

a. 15 cm ____________
b. 30 cm ____________
c. 60 cm ____________
d. 1.0 m ____________
e. 1.5 m ____________
f. 2.0 m ____________
g. 2.5 m ____________
h. 3.0 m ____________

MAGNETIC FIELD AROUND A CURRENT-CARRYING CONDUCTOR

In this portion of the lab activity we will use Ampère's rule apparatus to test the effect of current in a conductor on the magnetic field.

U Position the small compasses (included with the apparatus) around the vertical conductor and place the small magnetic compass under the horizontal conductor. Turn the entire apparatus so that the compass needle under the horizontal conductor is parallel to the de-energized conductor.

5. If the conductor is not energized, what is the compass needle aligning with in Step U?

____

V Connect the vertical post of the apparatus to the negative terminal of a DC power supply with a lead wire. Then connect the horizontal post to the positive terminal of a DC power supply with a lead wire.

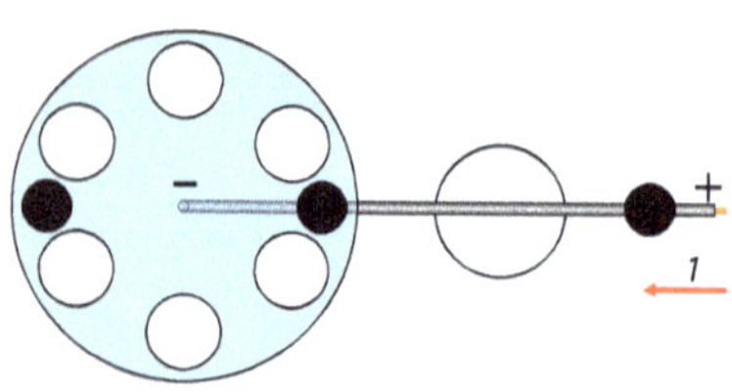

W Turn on the DC power supply and adjust the voltage to approximately 6.0 V. Observe the movement of the compasses. In the diagram at left draw in the orientation of the compasses by sketching in the needles and labeling the north ends. Turn off the power supply.

The right-hand rule for conductors is a mnemonic device in which you orient the thumb of your right hand in the direction of conventional current flow through a conductor, resulting in your fingers naturally curling around the conductor in the direction of the magnetic field lines.

6. Do your observations agree with this rule?

____

____

____

Name ______________________

7. What do you think would be the effect of reversing the current in the conductor?

____

____

____

X Reverse the leads to the apparatus. Turn the power supply back on. Observe the movement of the compasses. In the diagram at right draw in the orientation of the compasses by sketching in the needles and labeling the north ends.

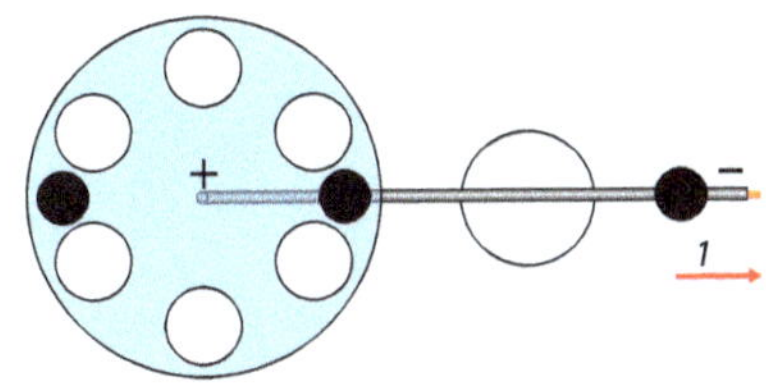

8. Did the magnetic field change as you predicted in Question 7? Explain.

____

____

____

____

____

Analysis

9. In the magnetic field diagrams, where were the field lines bunched together most closely relative to the magnet?

____

10. What happens to the magnetic field lines that run off Paper 1?

____

____

11. Is it possible for magnetic field lines to cross? Explain.

____

____

____

12. Explain how a compass would align itself if it were placed directly above the midpoint of a bar magnet?

____

____

____

13. Describe the pattern of field lines between opposite magnetic poles.

____

____

____

____

14. Describe the pattern of field lines between like magnetic poles.

15. Explain your observations when you checked for outside magnetic influences. Why did the compass needle drift away from the magnet?

Conclusion

16. In which direction does Earth's magnetic field point directly over the magnetic North Pole?

Going Further

17. When could it be difficult to navigate using Earth's magnetic field?

LAB

19A

Name

Date

Make the Wheels Go Round

Measuring Work in an Electric Motor

Electric motors have been in use for almost two centuries, performing numerous functions. The earliest electric motors were used for driving manufacturing equipment, mine ventilators, irrigation pumps, and elevators. Over time more and more electric motors entered our homes, converting electrical energy into other forms in blenders, refrigerators, fans, vacuums, and many more appliances. Today one of the areas of tremendous research in electric motors is in the transportation industry.

Tesla, Inc. has taken the lead in research and development of electric vehicles. In December of 2020 the Tesla Model 3 became the best-selling electric car in history—strong evidence that this twenty-year-old company is an industry leader. So why is there so much interest in electric vehicles? Some would argue that they are a cleaner form of transportation. And looking at just the operation of the vehicles, it is hard to argue against this point. But if we factor in the production of the electricity required to power them, the claim is less strong. Other people would argue that electric vehicles can be powered by non-fossil-fuel resources, and since fossil fuels are nonrenewable, a new technology will eventually be needed. Another group will argue for the efficiency of electric motors. We will leave the clean energy and alternative energy questions for another day.

In this activity we will investigate the efficiency of electric motors. We will use a simple DC motor to perform mechanical work and then compare the mechanical work done with the electrical energy consumed.

Does the load on an electric motor affect its efficiency?

QUESTIONS

» How much electrical energy does a DC motor consume to do mechanical work?

» How can I determine the efficiency of a DC motor?

EQUIPMENT

- multimeters (2)
- DC power supply
- DC electric motor
- support stand
- flask clamp
- wire leads with alligator clips
- lab rheostat, 0–170 Ω
- knife switch
- string
- spool
- metric mass set

1. What is the formula for calculating the mechanical work done when lifting an object?

2. What is power? What is the formula for calculating power?

3. What is the formula for calculating the electric power consumed in a circuit?

4. After completing this lab activity, a student determines that a motor performed 0.638 J of mechanical work while lifting a 50.0 g mass 1.30 m. The electrical energy delivered to the motor was 0.950 J. Calculate the efficiency of the motor using

$$\eta = \left(\frac{W_{\mathrm{M}}}{E_{\mathrm{E}}}\right)100\%.$$

Experimental setup

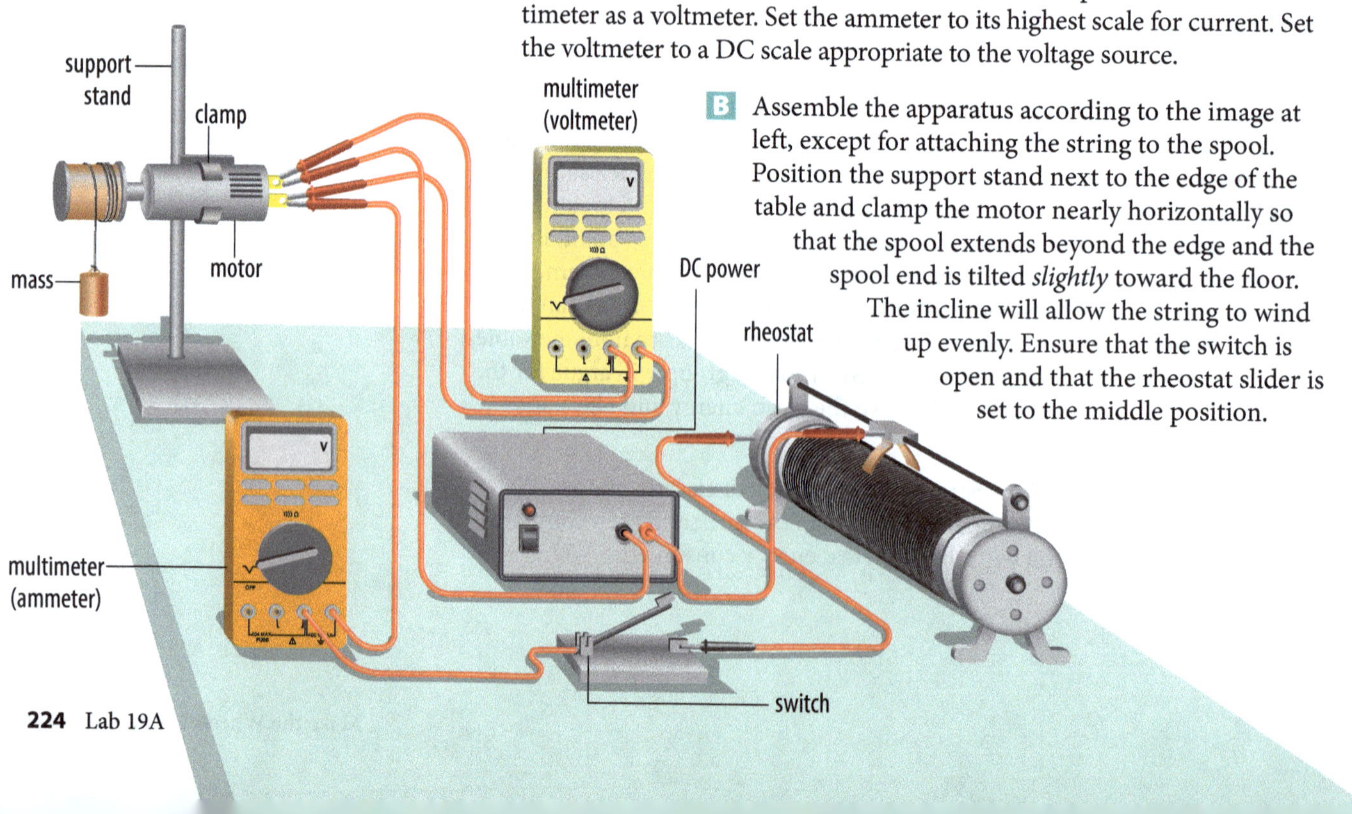

Procedure

A Set up one of the multimeters as an ammeter. Set up the second multimeter as a voltmeter. Set the ammeter to its highest scale for current. Set the voltmeter to a DC scale appropriate to the voltage source.

B Assemble the apparatus according to the image at left, except for attaching the string to the spool. Position the support stand next to the edge of the table and clamp the motor nearly horizontally so that the spool extends beyond the edge and the spool end is tilted *slightly* toward the floor. The incline will allow the string to wind up evenly. Ensure that the switch is open and that the rheostat slider is set to the middle position.

C Tie the string to the higher end of the spool. If the string slips, place a spot of white glue on the string to fasten it to the spool. The string should be 10–15 cm longer than the distance to the floor. Tie a loop in the end of the string. Start the motor and wind the string evenly onto the spool. Stop the motor when the string is completely wound onto the spool.

D Hook a 10 g mass onto the string and allow the mass to slowly descend to the floor. If the mass will not descend, use a larger mass. Rotate the spool to take up the slack in the string. Measure the height that the motor will lift the mass—the distance from the top of the mass's hook to the bottom of the spool. Record this height in Table 1.

E Close the switch, set the voltage on the power supply to approximately 1.5 V, and adjust the rheostat as necessary in order to obtain a constant motor speed. Open the switch just as the mass touches the spool. Allow the mass to descend and then close the switch again, repeating the mass lift until the mass rises at a constant speed. A second group member should practice starting and stopping the stopwatch when the motor is started and stopped. The switch operator should audibly identify these events to assist the individual who is timing. A third group member should read the two multimeters, adjusting the scales as needed to obtain appropriate readings. If a negative reading is obtained, swap the probes. The displays should be stable or at least not be changing rapidly while the motor is running.

F When the motor speed has been adjusted correctly, the individual with the stopwatch has practiced measuring the time interval, and the multimeters are set properly, you are ready to collect data. Close the switch and measure the time for the mass to be lifted from the floor to the spool. Record the time, motor terminal voltage, and the current through the circuit in Trial 1 in Table 1.

G Perform four additional trials with the 10 g mass according to Step F and record the results in Table 1.

H After collecting the data for the 10 g mass, turn off the power supply and immediately disconnect the motor from the circuit. Reconfigure the multimeter that was measuring voltage as an ohmmeter and measure the resistance through the motor. Record the resistance in Table 1. When finished, reconnect the motor to the circuit.

5. Why do we measure the motor resistance *after* the trials?

____

____

____

____

I Repeat Steps D–H with both a 20.0 g mass and a 50.0 g mass. The motor speed will have to be readjusted for the larger masses. The motor speed can be adjusted using the rheostat or the supply voltage. Record the data for Trials 6–10 with the 20.0 g mass and Trials 11–15 with the 50.0 g mass in Table 1.

Analysis

J Calculate the averages for lift time, motor terminal voltage, and circuit current for each set of five trials and record in Table 1.

K Assuming that the masses are accurate to 0.1 g, calculate the mechanical work done to lift each mass. Record your solutions in Table 2. Show your work for the 10.0 g mass in the space below.

L For each mass use the average currents and the motor resistances with the formula $V = IR$ to calculate the voltage drop in the motor windings. Record your solutions in Table 2. Show your work for the 10.0 g mass in the space below.

M Use the average current, average motor terminal voltage, and the motor voltage drop with $P_m = \bar{I}_m(\bar{V}_t - V_m)$ to calculate the power produced by the motor. Record your solutions in Table 2. Show your work for the 10.0 g mass in the space below.

Name

N Calculate the electrical work accomplished by the motor using the power produced by the motor and the average time, using $W_m = P_m\overline{\Delta t}$. Record your solutions in Table 2. Show your work for the 10.0 g mass in the space below.

O Determine the percent difference between mechanical work done to lift each mass and the electrical work done by the motor. Record your solutions in Table 2. Show your work for the 10.0 g mass in the space below.

P Using the formula $P_E = \overline{V}_t\overline{I}_m$, calculate the electric power delivered to the motor by the circuit by finding the product of the average value for motor terminal voltage and average circuit current. Record your solutions in Table 2. Show your work for the 10.0 g mass in the space below.

Q Calculate the electrical energy delivered by the circuit to the motor by multiplying the electric power delivered by the circuit by the average time interval for lifting each mass, using the formula $E_E = P_E\overline{\Delta t}$. Record your solutions in Table 2. Show your work for the 10.0 g mass in the space below.

R Finally, compare the mechanical work done by the motor with the electrical energy supplied to the motor by calculating the efficiency of the motor for each mass using

$$\eta = \left(\frac{W_M}{E_E}\right)100\%.$$

Record your solutions in Table 2. Show your work for the 10.0 g mass in the space below.

6. What was the average efficiency of the motor used by your lab group? According to your data, how is efficiency related to the size of the load on a motor?

___

___

___

___

Name ______________________

Conclusion

7. Explain why a motor turns faster as more current is supplied to it. Include in your discussion appropriate references to electrodynamic theory.

8. Discuss three changes that would make the motor more efficient.

Going Further

9. Research the efficiencies of gasoline car engines. Does your research and the efficiencies that you achieved in this activity support a claim that electric vehicles are more efficient than gasoline engines?

10. To fully evaluate the efficiency of electric vehicles, we must consider the energy transformations needed to produce the electricity. Describe the process for producing electricity from natural gas, which is the most commonly used fuel in the United States for electricity production. Include the types of energy involved in each step.

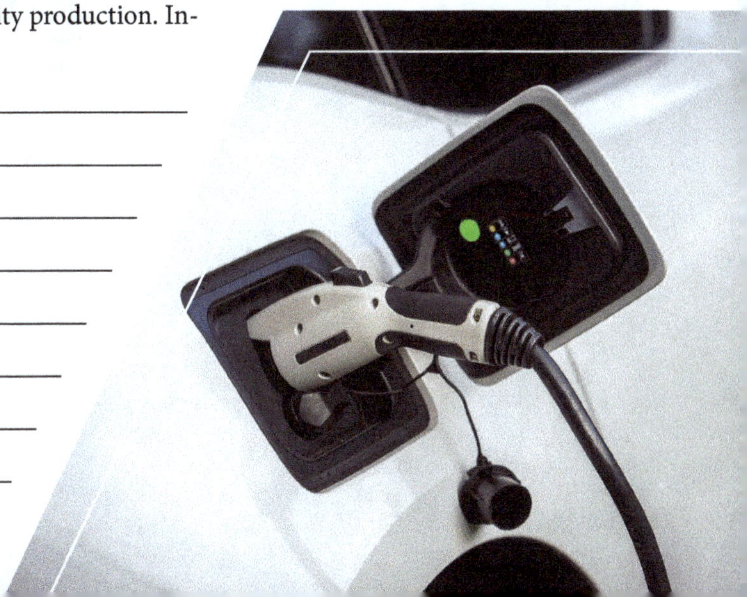

11. What are potential drawbacks to electric vehicles?

____

____

____

____

____

TABLE 1 *Data*

10.0 g Mass

Motor Resistance (Ω)

Trial	Lift Height (m)	Lift Time (s)	Motor Terminal Voltage (V)	Circuit Current (A)
1				
2				
3				
4				
5				
Avg.				

20.0 g Mass

Motor Resistance (Ω)

Trial	Lift Height (m)	Lift Time (s)	Motor Terminal Voltage (V)	Circuit Current (A)
6				
7				
8				
9				
10				
Avg.				

Name ______________________

50.0 g Mass

Motor Resistance (Ω)

Trial	Lift Height (m)	Lift Time (s)	Motor Terminal Voltage (V)	Circuit Current (A)
11				
12				
13				
14				
15				
Avg.				

TABLE 2 *Calculations*

Calculations

Mass (g)	Mechanical Work (J)	Motor Voltage Drop (V)	Motor Power (W)	Motor Work (J)	Percent Difference	Electric Power (W)	Electrical Energy (J)	Efficiency (%)
10.0 g								
20.0 g								
50.0 g								

LAB

19B

Name

Date

Maglev Training

Designing a Maglev Train

The transportation industry is always innovating. The goal is to move more cargo and more people greater distances while doing it more quickly and efficiently. One of the greatest impediments to efficiency is friction. Engineers are always looking for ways to reduce frictional losses. Railway engineers know that getting a train off its track is a great way to reduce friction—if they can lift the train. But how can we lift an entire train off its tracks?

One approach that engineers are considering is magnetic levitation, or maglev. Engineers realize that if they face the like poles of two magnets toward each other, they will repel each other. If there is enough magnetic force, they may be able to lift an entire train using that property. They could then use other magnets to propel the train forward. The train could theoretically never come in contact with the track that it is "riding" on.

In this STEM activity you will concentrate on just the lifting portion of the system. You will have significant freedom in choosing your materials, but remember that your electromagnet maglev system must lift your train, so the more material you use to build the train, the more train you will have to lift!

How can I lift a train with magnetic force?

QUESTIONS

» How can I design a maglev train?

» How can I produce enough magnetic force to lift my train?

EQUIPMENT

- maglev track

Procedure

DESIGN PARAMETERS

As you design your train, you must work within the design parameters listed below.

- Your train must fit the track designated by your teacher. The key measurement of the track is the width.

- Your train's electromagnetic maglev lift system must be able to lift the train.

Success for your train will be indicated by accomplishing two tasks: Your train (1) must be able to hold itself off the track and (2) should be able to travel the entire length of the track when propelled.

PLANNING THE DESIGN

A Conduct research on maglev and electromagnets.

B Keeping in mind the parameters listed above, brainstorm possible designs for your train with your team.

C Draw and label a diagram of your plan for your train. Indicate that this is a *preliminary design*.

D Share your research findings with your team. Reach a consensus on the final design that your group will build and submit for testing.

E Have someone in the group draw and label a diagram of the final design in the same manner as was done in Step C. Indicate that this is an *initial production design*.

BUILDING THE DESIGN

F Construct your train according to your initial production design.

G Conduct testing on your electromagnet maglev system.

TESTING THE DESIGN

H Place your train at one end of the track.

I Start the maglev system and demonstrate that the train is completely suspended by the maglev system.

J Propel the train toward the other end of the track as directed by your teacher and record the distance that it travels. ***To avoid overheating, leave your electromagnet powered only long enough to run the test.***

REFINING THE DESIGN

K On the basis of your testing results, make modifications to your train, if needed. If changes are made, draw and label a new diagram, indicating that this is the *final production design*. Include a description of how any proposed changes to the original design are intended to improve the performance of the new design.

RETESTING THE DESIGN

L If time permits, update your train and repeat Steps H–J to retest your modified design.

M Submit your preliminary, initial production, and final production designs as well as your analysis regarding the performance of your train.

LAB
20A

Name ____________________

Date ____________________

Illuminating Illumination

Investigating Luminous Flux and Illuminance

Taking photographs today is fairly simple with our cellphone cameras. But to get top-quality photographs, professional photographers use cameras that require them to adjust the settings for correct exposure. In the day of film cameras photographers would first select film of a particular speed. They would then consider the amount of available light so that they could determine the exposure time and the f-stop that they should use. F-stop is the ratio of the camera's focal length to its aperture.

Some photo shoots allow a photographer to add to the available light by using flood lights and camera flashes. Other applications allow for the use of ambient light only. A photographer needs to understand the relationships between the light available from the light source, the distance to the subject, the reflective nature of the subject, and the settings of his camera. In this activity we will investigate how illuminance changes due to the luminous flux of the source and to variations in distance from the light source.

How do professional photographers get their lighting right?

QUESTIONS

- How does changing the distance to the light source change the illuminance of a surface?
- How does changing the energy emitted from a light source change illuminance?
- How can I solve real-world illuminance problems?

EQUIPMENT

- multimeter
- meter sticks (2)
- stands
- lamp
- clear light bulbs (25, 50, 75, and 100 W)
- photovoltaic cell
- lead wires with alligator clip (2)
- hot mitts (1 pair)

1. Define each of the following and give its SI unit and symbol.
   a. luminous flux
   b. luminous intensity
   c. illuminance

Procedure

ILLUMINANCE (CURRENT) VERSUS DISTANCE

2. What do you expect to be the relationship between illuminance and the distance from a light source?

The first portion of the activity will investigate how changing the distance between the light source and the sensor affects the illuminance. You will do this by measuring how much current a photovoltaic cell can produce by converting the incident light energy into electrical energy.

A Prepare the multimeter to act as a DC ammeter. Set the multimeter to the smallest measurement range.

B Set up the apparatus as shown at left. Install the 50 W light bulb in the lamp. Align the lamp so that the filament ends are aligned with the photovoltaic cell to approximate a point light source. Connect the multimeter to the wires from the photovoltaic cell with the alligator clips.

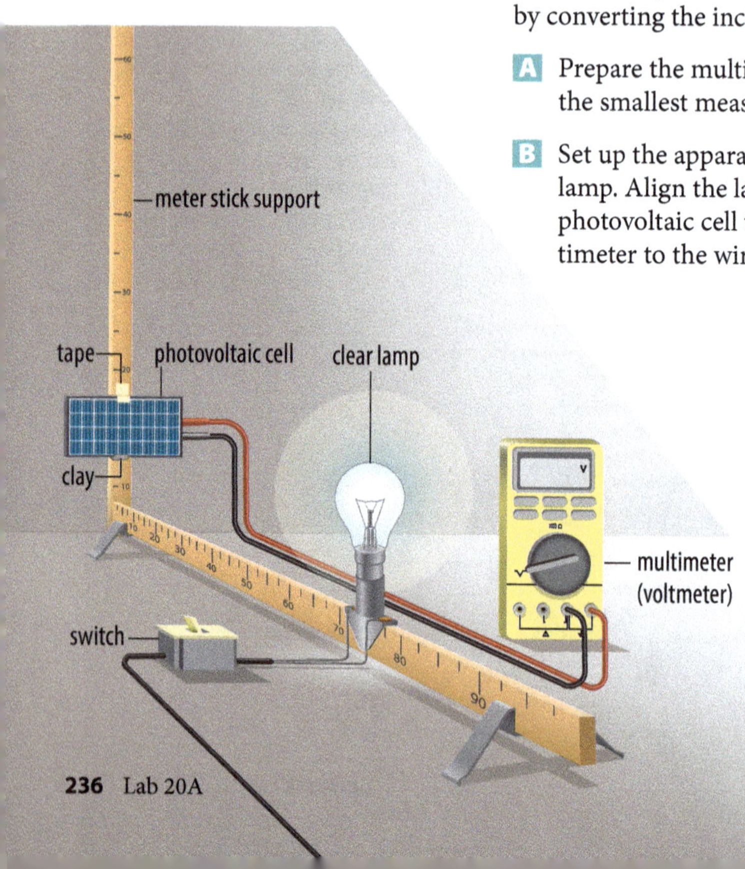

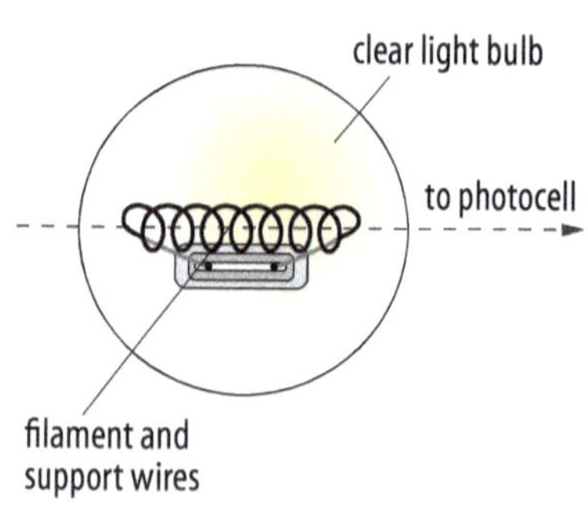

C Position the lamp so that the center of the filament is 15.0 cm from the photovoltaic cell.

D With the lamp switched off, measure the current produced by the photovoltaic cell. Record this ambient current in Trial 1 of Table 1.

3. Why do we measure this ambient current?

____

____

____

E Switch on the lamp and measure the current generated by the photovoltaic cell. Record this current in Trial 1 of Table 1. ***After the light bulb has been turned on, it will be hot. Avoid touching it with your bare hands!***

F Turn off the lamp.

G Repeat Steps C–F for Trials 2–6, increasing the distance 15.0 cm for each trial.

ILLUMINANCE (CURRENT) VERSUS LUMINOUS FLUX

For this series of trials you will measure the difference in illuminance at a fixed distance for light sources of different luminous intensities.

4. What do you expect the relationship between illuminance and luminous flux to be? Explain.

____

____

____

____

____

H Move the lamp so that the center of the filament is 30.0 cm from the photovoltaic cell.

I With the lamp switched off, measure the current produced by the photovoltaic cell. Record this ambient current in Trial 1 of Table 2.

J Switch on the lamp and measure the multimeter current and record in Table 2.

K Turn off the lamp.

L Repeat Steps I–K for three additional trials, replacing the bulb with the 25, 75, and 100 W bulbs for each subsequent trial.

Analysis

ILLUMINANCE (CURRENT) VERSUS DISTANCE

M Subtract the ambient current from the current to compute the corrected current for each trial in Tables 1 and 2.

N Create a scatterplot of the corrected current and source distance using the data from Table 1. As always, include a curve of best fit and the r^2 value of the fit.

5. Describe the shape of the graph and explain what it indicates. Include the equation of the best-fit curve and its r^2 value.

O Compute 1/source distance for each trial and enter your results in Table 1.

P You will likely find that the scatterplot of corrected current-versus-source distance looks like an inverse function. To check this, create a scatterplot of corrected current versus source distance. If your scatterplot has a linear fit, your claim of an inverse function is supported. Create this scatterplot using the data from Table 1, including the curve of best fit and the r^2 value of the fit.

6. Describe the shape of the graph and explain what it indicates. Include the equation of the best-fit curve and its r^2 value.

Q You will likely find that the scatterplot of corrected current versus 1/source distance will look like a quadratic function. To check this, create a third scatterplot of corrected current versus the square of 1/source distance. If this new scatterplot has a linear fit, your claim of an inverse function is supported. Create this scatterplot using the data from Table 1, including the curve of best fit and the r^2 value of the fit.

7. Describe the shape of the graph and explain what it indicates. Include the equation of the best-fit curve and its r^2 value.

8. Does the data support your hypothesis from Question 2? Explain.

Name ______________________

ILLUMINANCE (CURRENT) VERSUS LUMINOUS FLUX

R Using the assumed luminous intensity for each bulb given in Table 2, calculate the luminous flux for each bulb. Recall that $\Phi_v = 4\pi I_v$. Record these values in Table 2.

S Create a scatterplot of corrected current versus luminous flux for the data in Table 2.

9. Describe the shape of the graph and explain what it indicates about the relationship between luminous flux and illuminance. Include the equation of the best-fit curve and its r^2 value.

____

____

____

10. Does the data support your hypothesis from Question 4? Explain.

____

____

____

Conclusion

11. Could we have taken just one measurement of the ambient current? What would be the benefits and drawbacks to taking this approach?

____

____

____

____

____

____

____

____

____

12. How does your data compare to the equation for illuminance (Equation 20.1 from your textbook)? Explain.

____

____

____

____

Going Further

13. A photographer has a single 1455 lm light source that is 4.25 m from her subject. Her meter readings are indicating that she has one-third of the light needed for her photograph. How should she move the light to account for this? Explain.

TABLE 1 *Distance vs. Illuminance Data*

Trial	Source Distance (m)	Ambient Current (mA)	Current (mA)	Corrected Current (mA)	1/Distance to Source (m^{-1})	1/Distance to Source Squared (m^{-2})
1	0.150					
2	0.300					
3	0.450					
4	0.600					
5	0.750					
6	0.900					

TABLE 2 *Luminous Flux vs. Illuminance Data*

Trial	Source Power (W)	Ambient Current (mA)	Current (mA)	Corrected Current (mA)	Assumed Intensity (cd)	Luminous Flux (lm)
1	50				55.0	
2	25				21.0	
3	75				90.0	
4	100				125	

LAB

20B

Name ______________________

Date ______________________

Mirror, Mirror

Investigating Plane Mirrors

Checking out at the supermarket is pretty quick these days, but it wasn't always this way. In the not-so-distant past the cashier had to look for the printed price tag on each item and then manually enter the price into the cash register. Today most supermarkets have automated systems that scan an item's black and white barcode. The barcode communicates the price to a computer. The system works by shining a bright light on the barcode. The light reflects off the barcode, and through a series of mirrors and a rotating disk, the reflected light is converted into an electrical signal that uniquely identifies the product to the computer. A very experienced cashier can process a large order in a fraction of the time that it once took. In this lab activity we will investigate the behavior of light as it interacts with plane mirrors.

1. Describe the path of a light ray in a uniform medium.

____

2. State the law of reflection.

____

____

____

How does a computer use light to read a barcode?

QUESTIONS

- How can I use rays to locate the image produced by a plane mirror?
- How are the angles of incidence and reflection related?
- How do the measurements of an image compare to those of the object?
- What kind of image does a plane mirror produce?

EQUIPMENT

- plane mirror
- mirror support
- fine-point marker
- Drawing Area sheet (p. 246)
- corrugated cardboard
- sewing pins (round-headed)
- colored pencils (3, different colors)
- metric ruler
- protractor
- masking tape

3. Think about looking at yourself in the bathroom mirror. Where does your image seem to be located? Is it upright or inverted? Is it reversed left to right? How is your image's height compared to your height? How far is your image from the mirror compared to how far you are from the mirror?

Procedure

A Use a loop of masking tape to fasten the back of the mirror to the mirror support so that the mirror stands vertically.

B Using a fine point marker, place a mark in the center of the base of the mirror.

C Remove the Drawing Area sheet from the end of this lab activity.

D Tape the Drawing Area sheet to the cardboard at the corners.

E Place the mirror and its support on the paper so that the reflective surface coincides with the mirror line on the drawing area and the mirror center mark is on the long line perpendicular to the mirror (see below).

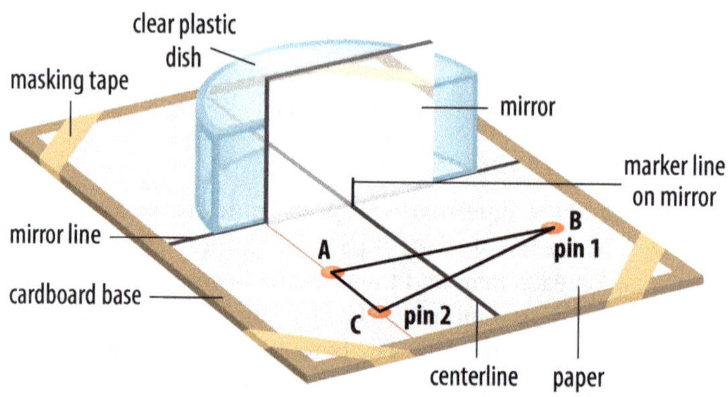

F Place a pin at Point A of ΔABC on the drawing area.

G Use one eye to look into the mirror from the opposite side of the centerline. Move your head left or right until the marker mark is lined up with the image of the pin located at Point A (which seems to be behind the mirror). Insert a second pin in line with your eye and the marker mark. This pin should cover the marker mark and the image of the pin in the mirror (see below). Label this Pin A_1 on the paper with one of the colored pencils.

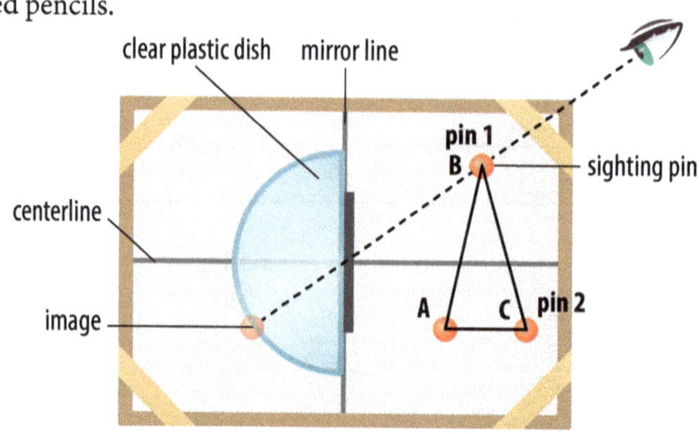

H Use one eye to look into the mirror from the same side of the centerline as the pin at Point *A*. Move left or right until the pin at Point *A* is lined up with its image in the mirror. Insert another pin in line with your eye and the pin at Point *A*. This pin should cover the pin at Point *A* and its image in the mirror (see below). Label this Pin A_2 on the paper with the same-colored pencil.

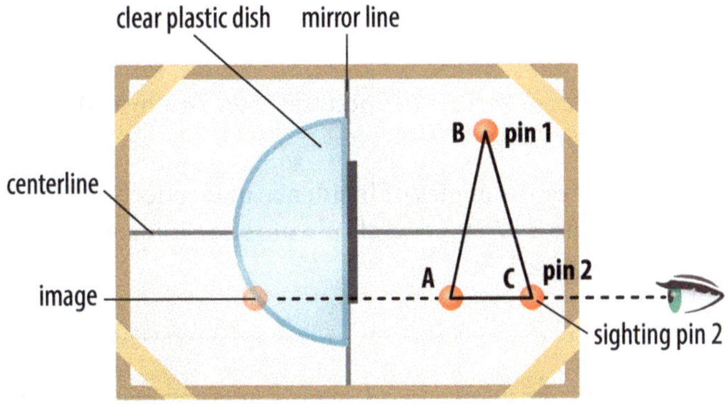

I Remove the pins from the paper.

J Repeat Steps F–I for Points *B* and *C* on the triangle using a different-colored pencil for each of the other two points. You should end up with pinned points A_1, A_2, B_1, B_2, C_1, and C_2 labeled on your paper.

K Remove the mirror and its support from the paper, and then carefully remove the Drawing Area sheet from the cardboard.

L Using a ruler, draw a line with the appropriately colored pencil from the hole left by Pin A_1 through the intersection of the two perpendicular reference lines, and extend the line all the way to the edge of the paper (see below). Label this Line a_1.

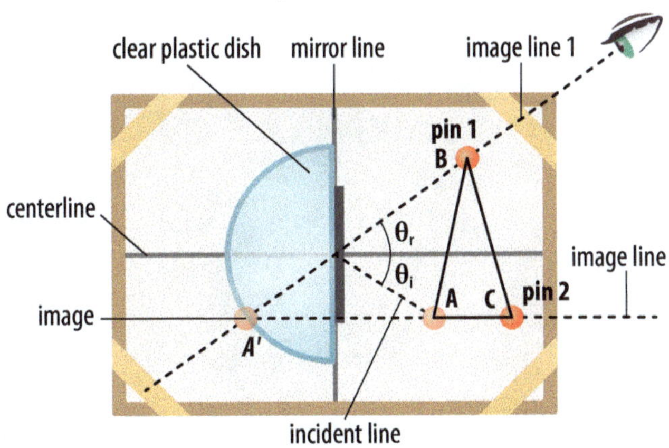

M With the same-colored pencil draw a second line from Point *A* on the triangle to the intersection of the two reference lines. Label this line a_3.

N With the same-colored pencil draw a line from the hole left by Pin A_2 through Point *A* of the triangle, and extend the line until it crosses image line a_1 (see image above). Label this line a_2. Label the intersection of lines a_1 and a_2 as Point A'.

O Repeat Steps L–N to locate Points B' and C' of the image.

Analysis

P Using the ruler, connect Points A', B', and C' to produce the image on the Drawing Area sheet.

4. How does your image compare to ΔABC? Explain.

____

____

Q Measure the angles of incidence and reflection and record the values in Table 1.

5. How do the sizes of the angles of incidence and reflection for a given reflected ray compare? What general rule of reflection could be suggested from this observation?

____

____

____

____

R Measure the image distance from the mirror line to each point on the image. Record this distance in Table 1.

S Calculate the percent differences between the triangle distances and the image distances for each trial. Record your results in Table 1. Show your work for Point A′ below.

6. How do the image and object distances compare for a plane mirror reflection?

____

____

The term *height* can refer to any lateral dimension of an object and its image. We will use it to refer to the sides of the triangles.

T Measure the image heights for Point A' (between A' and B'), Point B' (between B' and C'), Point C' (between C' and A') and record in Table 1.

U Calculate the percent differences between the triangle and the image heights for each point and record your results in Table 1.

7. How do the image and triangle heights compare for a plane mirror reflection?

____

____

Conclusion

8. On the basis of the experimental results, what can you conclude about the potential number of light rays emanating from any given point on an object? What aspect of your real-world experience with light reflecting off common objects gives assurance that this inference is correct?

____

____

____

____

9. What kind of images are produced by plane mirrors? State at least two evidences for your conclusion.

____

____

____

10. Do your experimental findings align with your common experience with plane mirrors? Explain.

____

____

____

Going Further

11. On the basis of what you learned in this lab activity, when the barcode scanner reads the barcode, do you think the number is reversed in the computer? Explain.

____

____

____

TABLE 1 *Data and Calculations*

Point	Angle of Incidence (°)	Angle of Reflection (°)	Triangle Distance (cm)	Image Distance (cm)	Percent Difference (%)	Triangle Height (cm)	Image Height (cm)	Percent Difference (%)
A			2.13			4.15		
B			3.48			3.30		
C			4.29			2.24		

Drawing Area 1

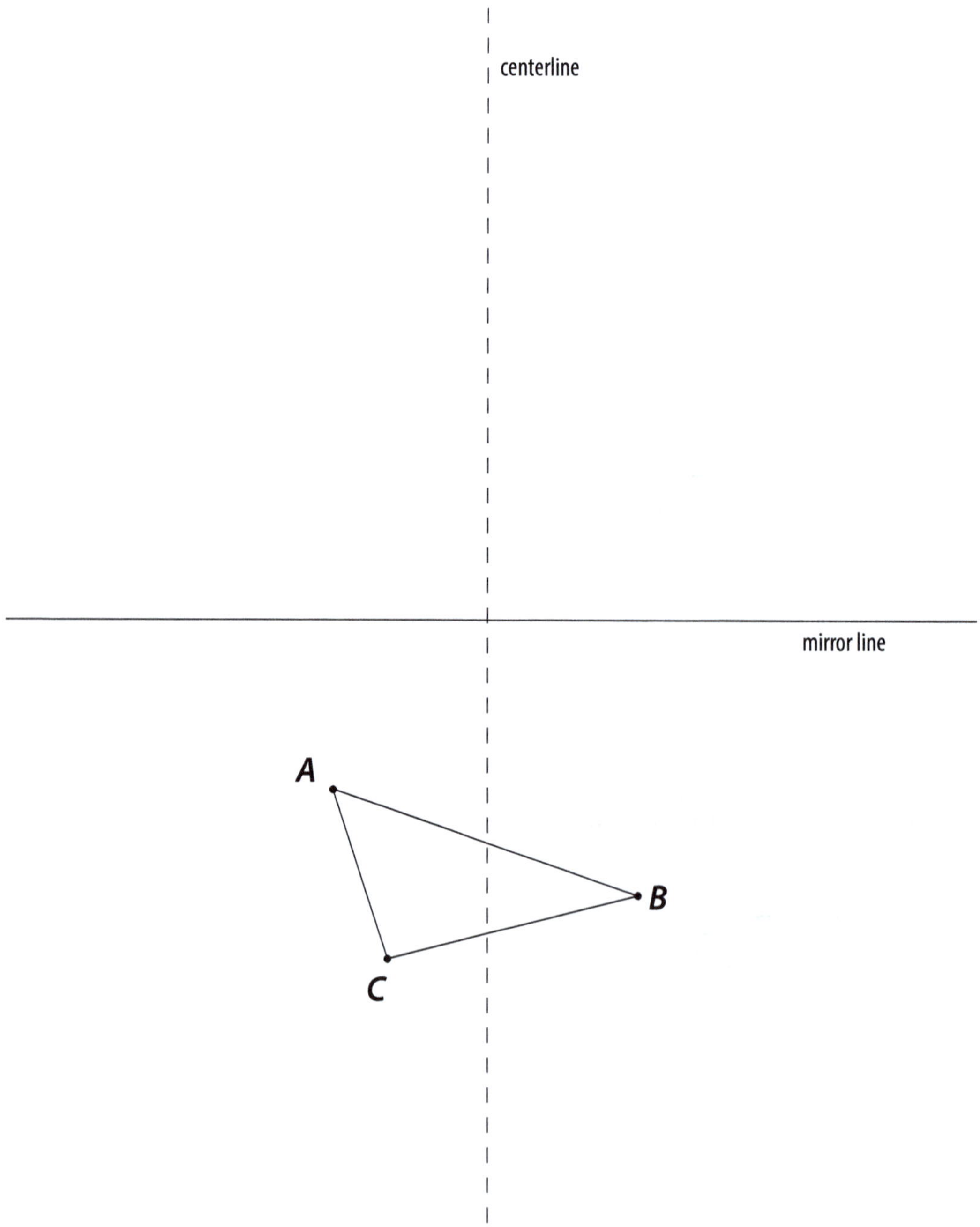

LAB

20C

Name ______________________

Date ______________________

Closer Than They Appear

Investigating Curved Mirrors

Most of us use mirrors every day. While plane mirrors are the most common, there are many instances in which we use curved mirrors. If you have ever used a makeup mirror, then you have used a curved mirror. Makeup mirrors use a concave mirror in what we consider Case 6, which produces an enlarged upright image. Convex mirrors are used in stores and hospitals. In stores convex security mirrors allow employees to monitor much of the store from a single vantage point. In hospitals convex safety mirrors allow people to see around corners so that they don't collide with a patient in a hospital bed or wheelchair.

When driving a car, we routinely check the passenger-side mirror—a convex mirror. How many of us, while sitting in the passenger seat, have wondered about the warning on the mirror that states that objects in the mirror are closer than they appear. Why is that? Is that always true? Does that reduce the helpfulness of that mirror? In this lab activity you will use a concave mirror to investigate how an image changes as an object is moved relative to the focal length and radius of curvature.

Why are objects closer than they appear in the passenger-side mirror?

QUESTIONS

- » How do images produced in a curved mirror differ from those produced by a plane mirror?
- » How can I use rays to locate the image produced by a curved mirror?
- » How can I determine the focal length of a curved mirror?
- » Does using distances to calculate magnifications produce different results from those using heights to calculate magnification?

EQUIPMENT

- concave mirror and support
- meter sticks (2)
- support stands
- focusing screen and support
- light source and support
- metric ruler
- computer paper (3 pcs.)

1. For a curved mirror, what do the center of curvature, the principal focus, and the vertex of the mirror have in common?

____

2. In what direction will a light ray that passes through the focal point of a spherical concave mirror be reflected?

____

3. Write the mirror equation and define each of the variables.

____

____

____

Procedure

DETERMINING FOCAL LENGTH

A Assemble the mirror, mirror support, meter stick, support stands, screen, and screen support according to the image below. The mirror should be positioned at one end of the meter stick with the concave surface facing the screen.

B Darken the room as much as is practical. Aim the apparatus at a bright, distinct object, preferably outside, at least 30 m away.

C Slowly move the screen back and forth along the meter stick until a sharp image is formed on the screen. You may have to realign the meter stick or the mirror in its holder several times until you find the image.

D Measure the distance from the center of the mirror to the screen and record the image distance in Table 1 for Trial 1.

E Move the screen so that the image is out of focus.

F Repeat Steps C–E four more times. Record these values in Table 1.

G Average the five values and enter the result in Table 1. Also enter the average as the image distance for Trial 1 in Table 2 and as the measured focal length in Table 3.

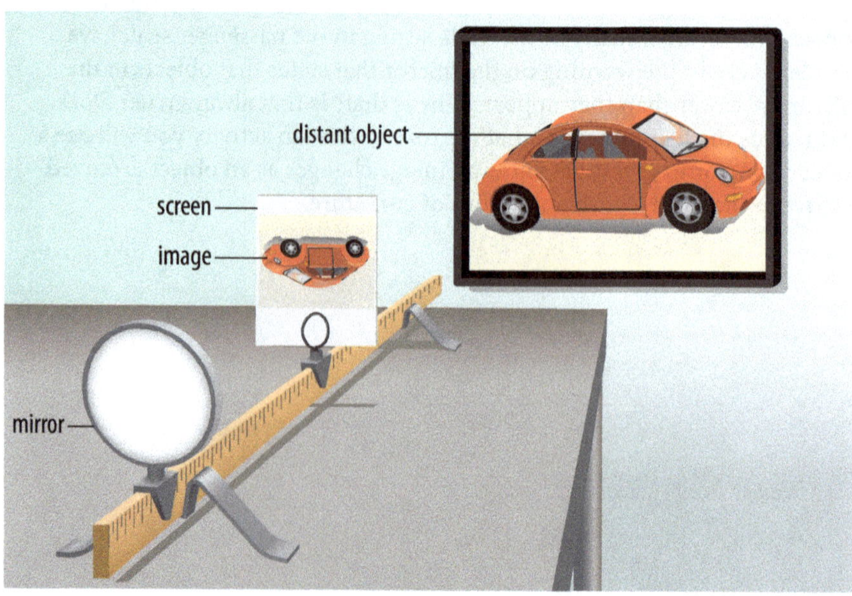

Name ______________________

4. We are claiming that this image distance is the focal length. Which case for convex mirrors is this? Explain.

____

____

____

____

5. Do you believe that this is a reasonable assumption? Explain.

____

____

____

____

____

H Compare the image to the object. Note the relative size of the image as well as its orientation (upright or inverted) and image type (real or virtual). Record these observations in Table 2.

CASES OF A CONCAVE MIRROR

I Calculate the approximate value of r from f. Recall that $f = r/2$, so $r = 2f$. Record this value in Table 2.

J Place a light source at a position somewhat greater than r from the mirror and just far enough from the side of the meter stick so that the focusing screen will not block illumination of the mirror (see below).

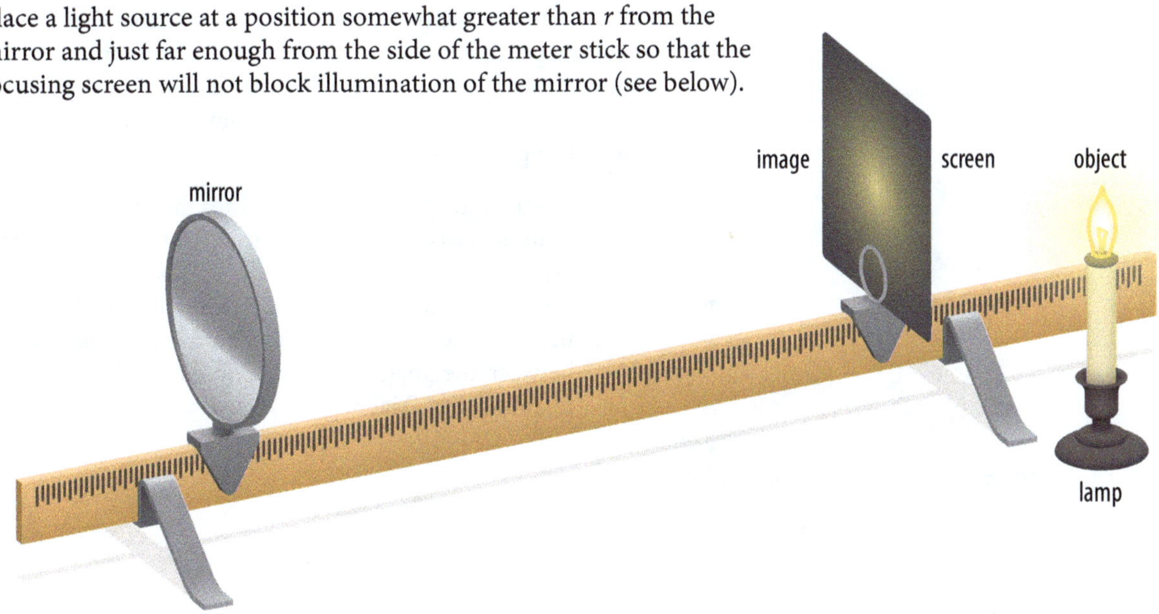

K Move the screen back and forth until a sharp image of the object is on the screen. Measure the distance between the object and the mirror as well as the distance between the image and the mirror. Record the object and image distances for Trial 2 in Table 2. Compare the image to the object. Note the relative size of the image as well as its orientation (upright or inverted) and image type (real or virtual). Record these observations in Table 2.

L If the image is bright enough, measure its height. Measure the corresponding height on the light source itself. Record the object and image heights in Table 2 for Trial 2.

M Move the light source to the distance *r* from the mirror. Repeat Steps K–L at this position and record your data in Table 2 for Trial 3.

N Place the light source at a position between *f* and *r* for Trial 4. Record object and image data according to Steps K–L in Table 2 for Trial 4.

O Place the light source between the mirror and *f* for Trial 5. You will find that you cannot form an image on the screen. To see the image, remove the screen and look into the mirror. For this trial the object should be positioned a few centimeters to the side of the apparatus.

P Tape three sheets of computer paper together end to end and draw a line down the center for the entire length of the paper. This line will represent the principal axis. Slide the paper under the apparatus so that it lines up with the centerline on the paper.

Q Mark the position of the object with an X directly beneath it. Mark the position of the mirror on the paper by making a short line across the centerline at the position of the mirror's vertex. Be sure that the center of the mirror is directly above the centerline drawn on the papers.

R Sketch two sighting lines to the same point on the image seen in the mirror. One method is to have your partner hold a pencil or thin dowel vertically near the mirror while you hold a second pencil vertically in line with the edge of the image and the pencil that your partner is holding. Mark both points on the paper and circle them. Label these points A_1 and A_2. Repeat the procedure from the other side of the centerline so that you have points for two rays pointing to the same point on the image. Label these two points B_1 and B_2. The image below depicts a method for finding these sighting lines using a board.

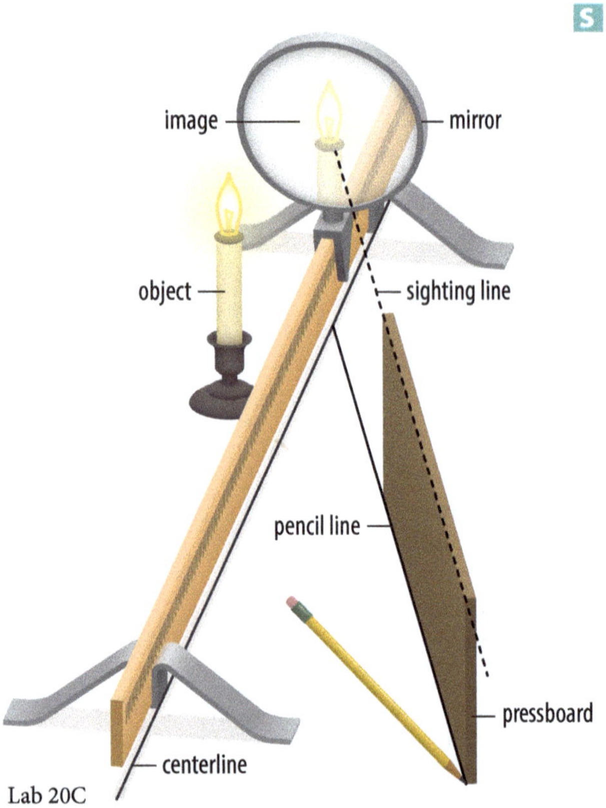

S Remove the paper and draw lines through both pairs of dots marked in Step R so that they intersect on the far side of the mirror position. This is the position of the virtual image from Trial 5. Measure the distance between the object and the mirror and the distance between the image and mirror, and record these values in Table 2 for Trial 5.

T Measure the distance between the object and the principal axis. Record this as the object height in Trial 5 of Table 2. Repeat for the image height. See image below.

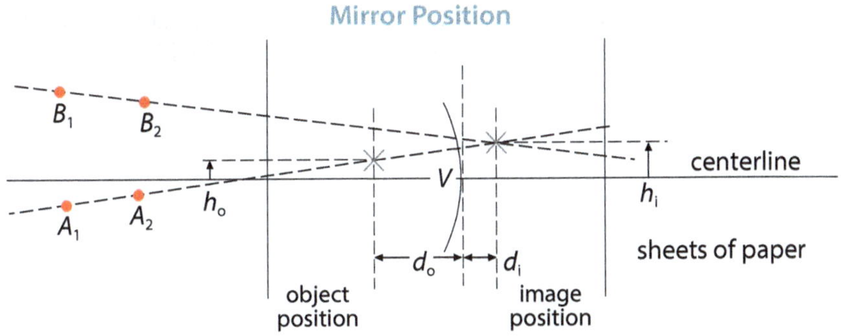

U Compare the image to the object. Note the relative size of the image as well as its orientation (upright or inverted) and type of image (real or virtual), and record these observations in Table 2.

Analysis

V For Trials 2–5 calculate the focal length. Record the results in Table 3. Show your work for Trial 2 below.

W Calculate the average of the calculated focal lengths and record your results in Table 3.

X Calculate the percent difference between the measured focal length and each calculated focal length, as well as the average calculated focal length. Record your results in Table 3. Show your work for Trial 2 below.

Y Calculate the magnification of the image for each trial using two different ways, using the ratio of image and object distances and then using the ratio of image and object heights. Record your results in Table 4. Show your work for Trial 2 below.

Z For Trials 2–5 compute the percent difference between the two methods and record your results in Table 4.

Conclusion

6. Consider the calculated focal length as compared with the measured focal length, assuming that the image of the object outside the window was at the focal length. State a claim as to whether this was a reasonable assumption.

7. How could you have made a better measurement of the focal length of the mirror?

Going Further

8. The passenger-side mirror of a car is a convex mirror. Explain why objects in the passenger-side mirror of a car are closer than they appear. What implication does this have when driving?

TABLE 1 *Direct Measurement of Focal Length*

Trial	Image Distance (cm)
1	
2	
3	
4	
5	
Avg. (f)	

TABLE 2 *Image Data*

Radius of Curvature (cm)

Trial	Object Position	Object Distance (cm)	Image Distance (cm)	Relative Image Size	Image Orien-tation	Image Type	Object Height (cm)	Image Height (cm)
1	$d_o = \infty$							
2	$d_o > r$							
3	$d_o = r$							
4	$f < d_o < r$							
5	$0 < d_o < f$							

TABLE 3 *Calculation of Focal Length*

Trial	Measured Focal Length (cm)	Calculated Focal Length (cm)	Percent Difference %
1			
2			
3			
4			
5			
Avg.			

TABLE 4 *Calculated Magnification*

Trial	Magnification (distances)	Magnification (heights)	Percent Difference %
1			
2			
3			
4			
5			

LAB

21A

Name ______________________

Date ______________________

All Bent Out of Shape

Investigating Refraction

Spearfishing is a method of fishing that uses a spear instead of a net or hook. Spearfishing has been around for thousands of years and is a popular sport today. It is often done by people swimming underwater, but it can also be done from the bank of a river or lake. Spearfishing from out of the water has the added dimension of needing to factor in the refraction of light.

Physicists describe the effect of a medium on the speed of light with the material's index of refraction, and they model the refraction of light using Snell's law. While a spearfisher may not understand indexes of refraction or Snell's law, he must understand refraction if he is to be successful. In this lab activity we will use data from the refraction of light to determine both the index of refraction for different media and the critical angle for an interface between two media.

1. What is refraction and why does it occur?

____

____

____

____

____

____

____

____

How does a spearfisher know where to aim?

QUESTIONS

» How can I determine the index of refraction for a material?

» How can I determine the critical angle for an interface?

EQUIPMENT

- protractor
- metric ruler
- semicircular dish
- corrugated cardboard
- plane mirror
- sewing pins, round-headed (3)
- computer paper (3 pcs.)
- masking tape

2. What is the index of refraction?

____

____

____

3. Write Snell's law and identify each variable.

____

____

____

____

Procedure

A Label the three pieces of paper 1, 2, and 3. Using the protractor and metric ruler, draw two perpendicular lines on each of the pieces of paper to divide the sheets into quarters. Label the long line "centerline" and the short line "interface."

B Inspect the semicircular dish and verify that there is an etch mark in the center of the flat side at the lower edge. If not, your teacher will give you alternate instructions for providing a reference mark at that location.

C Fill the dish about two-thirds full of water.

INDEX OF REFRACTION FOR WATER

The objective of this portion of the activity is to measure the angles of incidence and refraction as light travels from air into water. From this information you will determine the index of refraction for water. You will then compare the experimental values to the accepted value.

D Use the masking tape to fasten Sheet 1 at the corners to the cardboard base.

E Place the water-filled dish so that the flat side is on the interface line and the etch mark lies on the centerline (see below).

F Place a pin about 4 cm from the flat surface of the dish and about 10° to one side of the centerline. This pin will be referred to as the object. Label the pin position O_1.

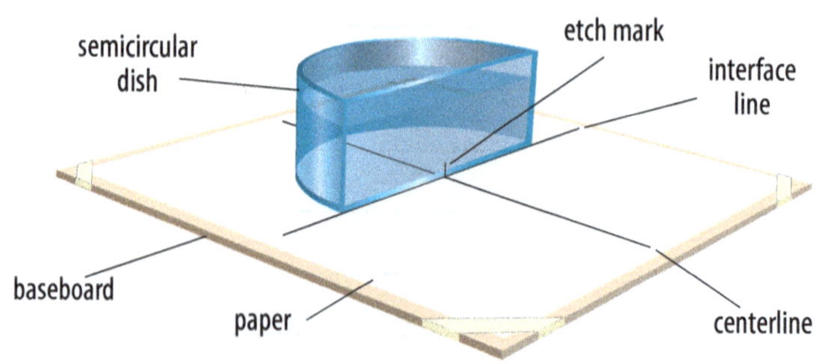

Name ______________________

4. Describe what will happen as a light ray passes from air and into the water at an angle. Explain why the light will behave as described.

____

____

____

____

G Set the entire apparatus near the edge of the table with the curved side of the dish facing the table edge. This arrangement will allow your eye to be at the same level as the dish. Close one eye and sight through the curved side of the dish to the object pin. Move left or right until you observe the object pin lined up with the etch mark in the dish. Insert a second pin on the curved side of the dish in line with the object and etch mark. This image pin should block the view of the etch mark and the object pin. Label the pin position I_1.

H Remove both pins.

I Repeat Steps F–H for three additional positions of the object pin. Place all three object pins in the same quadrant of the paper, increasing the angle from the centerline on each subsequent trial (see right). Label each of the object positions O_2, O_3, and O_4 and the corresponding image positions I_2, I_3, and I_4.

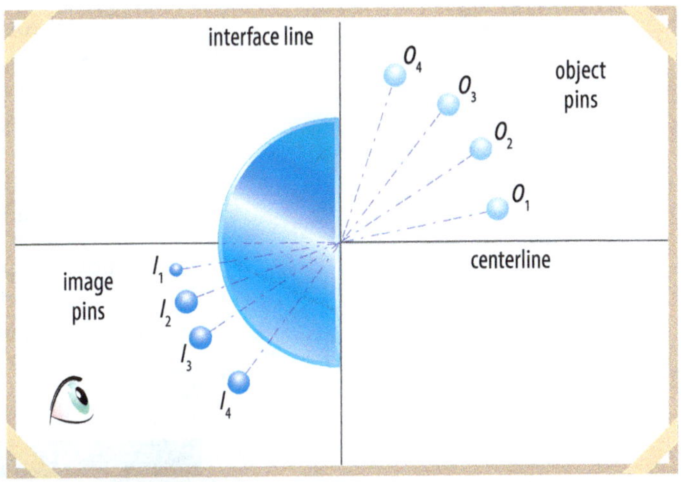

J Remove the dish from the paper and the paper from the cardboard. Connect each object pin position and each image pin position to the intersection of the interface line and centerline.

K Measure the angles of incidence and refraction for each trial and record your data in Table 1. The rays originate at *O* and travel to *I*.

CRITICAL ANGLE AT THE WATER-AIR INTERFACE

The objective of this portion of the experiment is to determine the critical angle as light travels from water into air. The concept of critical angle is important for the technology of fiber optics.

5. What is the critical angle? Under what conditions will an interface have a critical angle? Explain.

____

____

____

____

____

L Tape Sheet 2 to the cardboard base and place the water-filled dish on the interface as before. Again, set the apparatus at the edge of the table.

M Place a pin on the interface line at one end of the semicircular dish (Point X in the image below). Label this image pin position I.

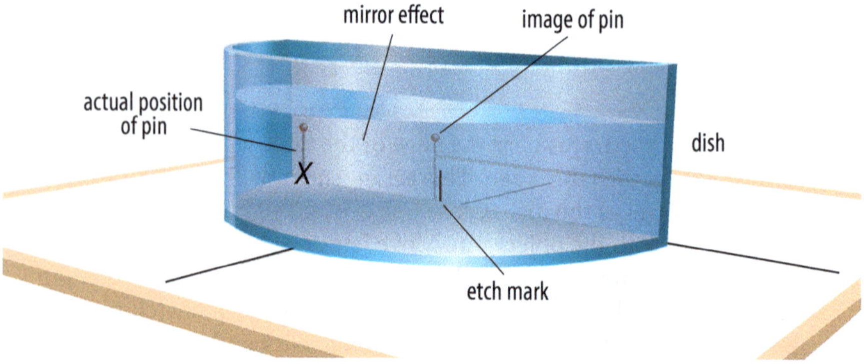

N Sight through the curved portion of the dish toward the etch mark and the quadrant that contains the image pin. Move your head from side to side until you can barely see the image pin (before the flat surface loses transparency). If you cannot see the pin, move it a few millimeters away from the interface line and repeat the sighting. Continue making small adjustments to the position of the image pin until it is just visible.

O Place a pin in line with the etch mark and the image pin on the curved surface of the dish. Label this pin position O for the object.

P Remove the pins and dish from the paper and remove the paper from the cardboard. Connect the object pin position to the intersection of the interface and the centerline.

Q Measure the critical angle for the object and record your result in Table 2.

INDEX OF REFRACTION FOR CROWN GLASS

If you place a mirror face-down on a flat surface, you will be able to look from edge to edge through the glass to the opposite side of the mirror. By using the mirror as you did the dish of water in the first experiment, you can determine the index of refraction in glass.

R Use the masking tape to fasten Sheet 3 to the cardboard base.

S Place the mirror on the paper so that one of the long edges is on the interface and the center of the mirror lies on the centerline (see left). Again, set the entire apparatus near the edge of the table with the mirror edge that is on the interface line away from the table edge.

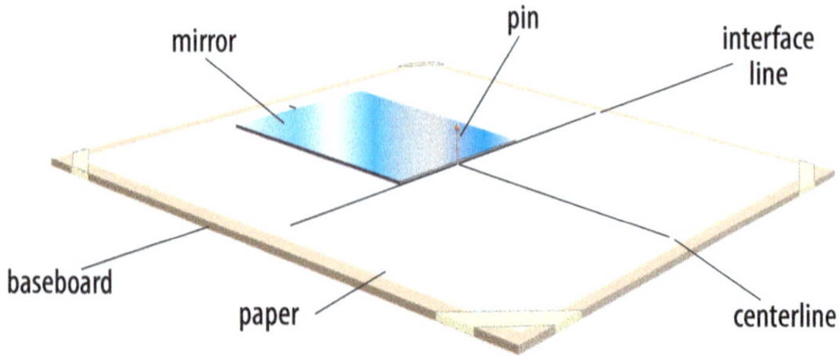

T Place a pin along the mirror's edge at the intersection of the interface and centerline to serve as the reference mark.

U Place a pin about 4 cm from the edge of the mirror that is aligned with the interface line and about 10° to one side of the centerline. This pin will be referred to as the object. Label the pin position O_1.

V Close one eye and sight through the edge of the mirror to the object pin. Move left or right until you observe the object pin lined up with the reference pin. Insert a pin right along the edge of the mirror on the side of the mirror through which you are sighting. This image pin should block the view of the reference and object pins. Label the pin position I_1.

W Remove the object and image pins.

X Repeat Steps U–W for three additional positions of the object pin. Place all three object pins in the same quadrant of the paper, increasing the angle from the centerline on each subsequent trial (see right). Label each of the object positions O_2, O_3, and O_4 and the corresponding image positions I_2, I_3, and I_4.

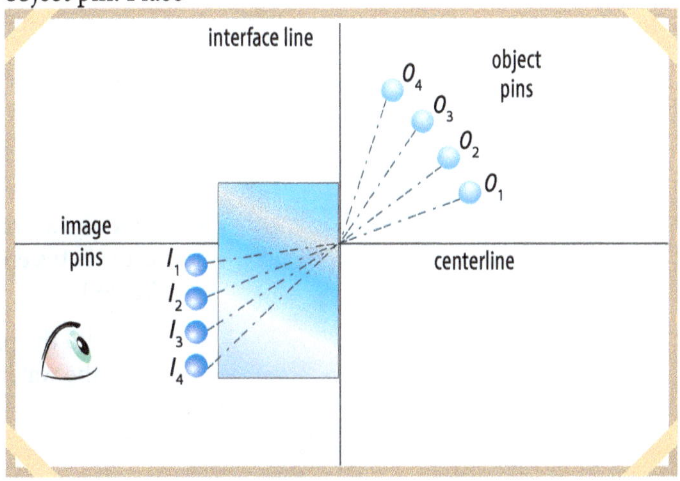

Y Remove the mirror from the paper and the paper from the cardboard. Connect each object pin position and each image pin position to the intersection of the interface line and centerline.

Z Measure the angles of incidence and refraction for each trial and record your data in Table 3. The rays originate at O and travel to I.

Analysis

INDEX OF REFRACTION FOR WATER

6. Calculate the index of refraction for water to the thousandths place for each trial in Table 1. Record your answers in Table 1. Show your work for Trial 1 below.

7. Calculate the percent error for the index of refraction for water in each trial of Table 1. Record your answers in Table 1. Show your work for Trial 1 below.

AA Find the average of the experimental values and determine the percent error between the average and the accepted value. Record these values in Table 1.

8. How did the procedures keep the light from being refracted as it moved from the water to the air on the curved side of the dish?

____

____

____

____

CRITICAL ANGLE AT THE WATER-AIR INTERFACE

9. Calculate the theoretical critical angle for a water-air interface. Record this value in Table 2.

Name ______________________

10. Determine the percent error for the critical angle. Record this value in Table 2.

INDEX OF REFRACTION FOR CROWN GLASS

11. Calculate the index of refraction for glass to the thousandths place for each trial in Table 3. Record your answers in Table 3. Show your work for Trial 1 below.

12. Calculate the percent error for the index of refraction for glass in each trial of Table 3. Use 1.523 as the accepted value. Record your answers in Table 3. Show your work for Trial 1 below.

Conclusion

13. Review the results of the first part of the experiment. Did you notice any pattern in the percent errors as the angle of incidence increased? If so, what may have caused this?

14. Would you anticipate the critical angle for crown glass to be larger or smaller than the critical angle for water? Justify your answer.

15. State a claim regarding the accuracy of your data.

16. According to your data, were your results for some of the experiments more accurate than others? If so, why do you think this was true?

Name ______________________

Going Further

17. You are trying your hand at spearfishing from the bank of a stream. Do you expect that the rays moving from a fish toward you will bend toward or away from the normal line as they pass through the water-air interface? Explain.

____

____

____

____

18. On the basis of your answer to Question 17, where should you aim in order to spear a fish?

____

____

____

____

Data

TABLE 1 *Index of Refraction for Water*

Trial	Index of Refraction, Air	Angle of Incidence (°)	Angle of Refraction (°)	Index of Refraction	Percent Error %
1	1.0003				
2					
3					
4					
Avg.					

TABLE 2 *Critical Angle*

Object	Measured Critical Angle (°)	Index of Refraction, Air	Index of Refraction, Water	Calculated Critical Angle (°)	Percent Error %
O_1		1.0003	1.333		

TABLE 3 *Index of Refraction for Crown Glass*

Trial	Index of Refraction, Air	Angle of Incidence (°)	Angle of Refraction (°)	Index of Refraction	Percent Error %
1	1.0003				
2					
3					
4					

LAB

21B

Name

Date

Call the Ball

Investigating Thin Lenses

The title of this lab activity may seem odd. The phrase *call the ball* is common in naval aviation and has a lot to do with lenses. The ball being referred to is the orange light of a Fresnel lens optical landing system (FLOLS). The FLOLS is a set of lights used to guide aircraft to a safe landing on aircraft carriers. The FLOLS consists of a bar of green lights that represent the horizon. Lenses cause the orange source light, or meatball, to move up or down relative to the horizon, indicating whether the aircraft's approach is on a proper glide slope. Flying too high may result in missing the carrier and ending up in the ocean, while flying too low could result in crashing into the back of the carrier. A pilot making an approach will be instructed to "call the ball" to indicate that he has the meatball in sight.

In this lab activity you will experiment with various object distances in order to determine the location and properties of the images formed by a lens. You will also determine the focal length of the lens by measuring object and image distances. Finally, you will determine the magnification of the images.

1. What is meant by a thin lens?

___

___

How does the device above help pilots land?

QUESTIONS

» How do the images formed by converging lenses change as I alter object distance?

» How can I measure the focal length of a lens?

» How do magnification values differ when using distances or height to calculate them?

2. Compare converging and diverging lenses.

3. When discussing lenses, how do the terms *convex* and *concave* relate to converging and diverging? Why are the two terms problematic when discussing lenses?

4. Write the thin-lens equation and define each of the variables.

EQUIPMENT

- converging lens
- meter sticks (2)
- support stands (2)
- focusing screen and support
- lens support
- lamp and support
- clear plastic protractor
- metric ruler
- clear tape
- computer paper (3 pcs.)

Procedure

MEASURING FOCAL LENGTH

A Darken the room as much as is practical.

B Tape a sheet of computer paper to a wall opposite an unshaded window. This will be the screen for this experiment. Hold the lens parallel to and about 1 m from the screen. Move the lens toward the screen surface slowly until a clear image of a distant object from outside the window is formed on the screen.

C Measure the distance from the screen to the plane of the lens. Record this image distance in Trial 1 of Table 1. Also record this as the measured focal length in Table 2.

D Describe the image, including relative size, orientation, and image type. Record these observations in Table 1.

5. Step C has you record the distance from the screen to the lens as the lens's focal length in Table 2, suggesting that this is Case 1 for converging lenses. Is this true? Explain.

Name ______________________

6. On the basis of your answer to Question 5, do you think it is valid to claim that the distance to the image is the focal length as Step C has you do? Explain.

____

____

____

____

CASES OF CONVERGING LENSES

E Assemble an optics bench with a lens, a lens support, a meter stick, support stands, a lamp, a lamp support, a screen, and a screen support according to the image at right. A clear plastic protractor should be positioned very close to the lamp facing the lens so that the light projects through the ruler scale. Position the protractor horizontally, with a major division mark at the lamp's centerline. The scale markings will show up clearly on the screen when the image is focused correctly. This will greatly aid making object and image distance measurements.

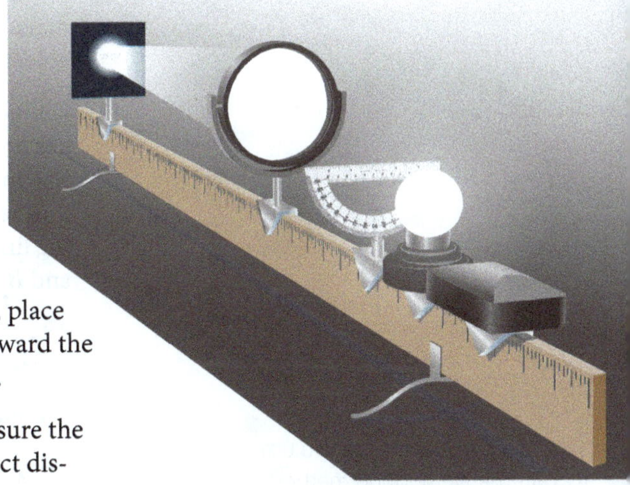

F Darken the room as much as is practical.

G With the lamp and protractor at the far end of the meter stick, place the lens just in front of the screen and slowly move the lens toward the lamp until the clearest possible image is formed on the screen.

H Using the meter stick that is not part of the optics bench, measure the distance from the center of the lens to the protractor (the object distance). Measure the distance from the center of the lens to the screen (the image distance). Record these distances in Table 1 for Trial 2.

I Describe the image, including relative size, orientation, and image type. Record these observations in Table 1.

J The height of the object can be obtained from the image. For example, if you can clearly see the 4 cm mark and the 6 cm mark from the protractor on the image, record 2.00 cm as the object height. Then, using the metric ruler, measure the distance between those same two points on the image and record this value as the image height. Record these heights in Table 1 for Trial 2.

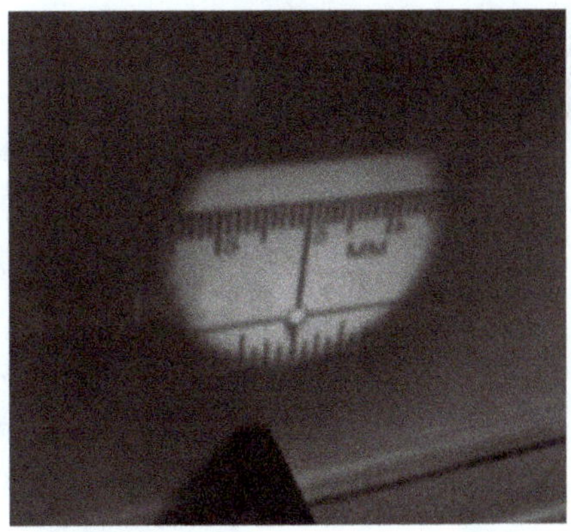

▲ Sharp image obtained using light projected through a protractor as the object

K Move the lens so that the distance between the lens and protractor is exactly $2f$ (image distance from Trial 1). Move the screen until the clearest possible image is formed. Repeat Steps H–J and record the data in Table 1 for Trial 3.

L Move the lamp toward the lens so that the distance between the protractor and center of the lens is between f (image distance from Trial 1) and $2f$. Again, move the screen until the clearest possible image is formed. Repeat Steps H–J and record the data in Table 1 for Trial 4.

M Reposition the lamp in the center of the meter stick and move the lens so that the distance between the protractor and lens is less than the focal length. (A distance of 10 cm is a good starting point.) Try to create a focused image by moving the screen closer to the lens. If you can obtain a focused image, move the lens closer to the lamp and try again until you are unable to produce a focused image on the screen.

N Remove the screen from the optics bench. Leave the lens and the object in place and move the optics bench so that you can look into the lens from the end where the screen had been.

O Tape three sheets of computer paper together end to end and draw a line down the center of the entire length of the paper. Slide the paper under the optics bench so that the bench is aligned with the centerline on the paper.

P Mark and label the positions of the object and the lens on the paper by making a short line across the centerline. Be sure that the center of the object and the center of the lens are directly above the centerline.

Q Define two sighting lines to a point on the image seen in the lens (a number on the protractor, for instance). Use two pencils to define a sighting line to the reference point. Mark the positions of the pencils A_1 and A_2 for the first sighting line. Reposition the pencils to define a second sighting line to the same point on the image. Mark these pencil positions B_1 and B_2 (see below).

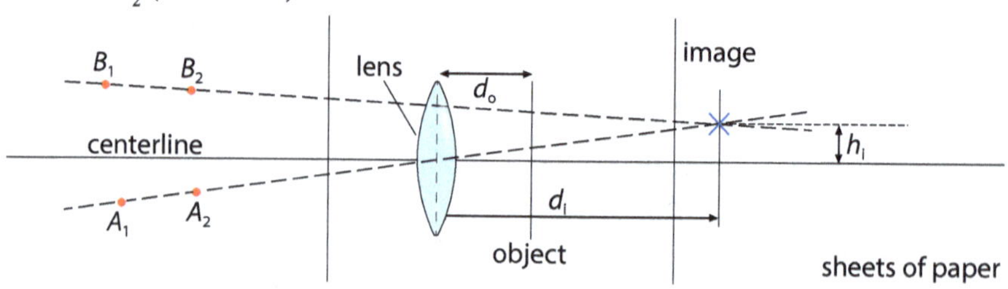

R Measure and record the distance from the protractor to the lens. Record this object distance in Table 1 for Trial 5.

S Measure the distance from the centerline of the protractor to the point that you used as the sighting mark in Step Q. Record this as the object height in Table 1 for Trial 5.

T Remove the paper strip from under the optics bench. Extend the sighting lines until they intersect. Measure the distance from the lens to the intersection of the sighting lines. Enter this image distance as a negative value in Table 1 for Trial 5.

U Draw a line from the image position (intersection of the sighting lines) perpendicular to the centerline. Measure the length of this perpendicular line as the image height and record it in Table 1 for Trial 5.

V Describe the image, including relative size, orientation, and image type. Record these observations in Table 1.

Your teacher might have you use this alternative sighting method using a thin pressboard.

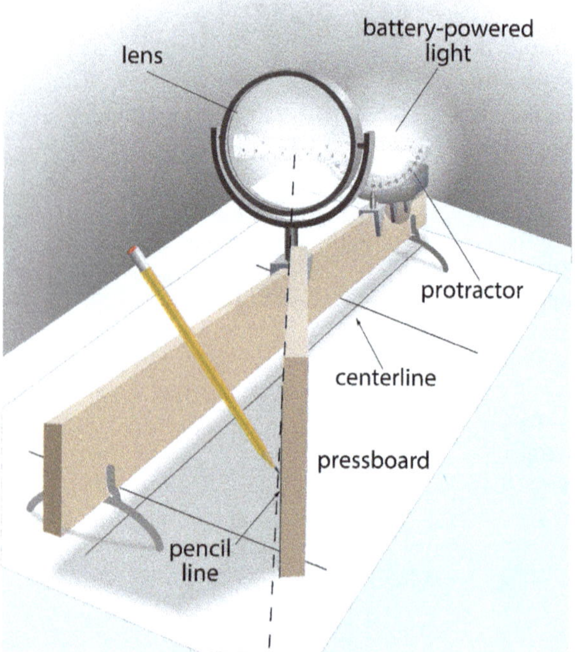

Name

Analysis

7. For Trials 2–5 use the object and image distances to calculate the focal length. Calculate the average of these calculated focal lengths and record your results in Table 2. Show your work for Trial 2 below.

8. Calculate the percent difference between the measured and calculated focal lengths for Trials 2–5. Record your results in Table 2. Show your work for Trial 2 below.

9. For Trials 2–5 calculate the magnification using both the ratio of distances as well as the ratio of heights. Record your results in Table 2. Show your work for Trial 2 below.

10. Calculate the percent difference between the two magnification values for Trials 2–5. Record your results in Table 2. Show your work for Trial 2 below.

11. Which cases for a converging lens do Trials 2–5 demonstrate? Which case is not represented and why?

Conclusion

12. Review your data and make a general statement that summarizes the relative size, location, and orientation of an image formed by a converging lens compared with the object as the object position changes from a distance greater than $2f$ from the lens to a location between $2f$ and f.

13. As the object approaches the lens from a distance less than f, what kind of image is produced?

14. On the basis of your observations, what would be the optimum viewing distance to obtain maximum magnification when using a converging lens?

15. Review your answer from Question 6. On the basis of your observations, do you think that using Trial 1 as Case 1 is valid? Explain.

16. Think about how image distance changes according to object distance. Do you think that the value obtained in Trial 1 for the focal length is too high or too low? Explain.

Going Further

17. What is the purpose of using lenses for optical landing systems such as the FLOLS?

18. Fresnel lenses are used in lighthouses to reduce the weight of the very large lenses that are used. Why do you think they are used in FLOLS, which have much smaller lenses?

TABLE 1 *Data*

Trial	Object Distance (cm)	Image Distance (cm)	Relative Image Size	Image Orientation	Image Type	Object Height (cm)	Image Height (cm)
1 ($d_o = \infty$)							
2 ($d_o > 2f$)							
3 ($d_o = 2f$)							
4 ($f < d_o < 2f$)							
5 ($0 < d_o < f$)							

TABLE 2 *Calculations*

Trial	Measured Focal Length (cm)	Calculated Focal Length (cm)	Focal Length Percent Difference %	M_d	M_h	Magnification Percent Difference %
1 ($d_o = \infty$)						
2 ($d_o > 2f$)						
3 ($d_o = 2f$)						
4 ($f < d_o < 2f$)						
5 ($0 < d_o < f$)						
Average						

LAB
22

Name

Date

In the Groove

Investigating Diffraction

Twenty years ago people in the music industry would have said that vinyl LP records were an artifact of history. The advent of the compact disk and today's solid-state storage devices seemed to signal the end of the LP record. These newer storage devices promised greater capacity as well as clearer playback sound. But today the LP market is booming as people long to hear that familiar hiss of an LP record.

Interestingly, music is stored on an LP in a way similar to the way that it is done on a CD. An LP has a single groove that loops around the LP, and during recording, small imperfections are etched along the groove. The needle on a record player interacts with the imperfections and generates vibrations that produce the sounds that come out of the speakers. A CD also has a spiral track etched in the plastic disk. A laser creates pits on a CD that record data. Another laser in the playback device reads the data recorded in the pits and converts that to sound. One advantage of a CD is its storage capacity. An LP typically has 8–12 songs recorded on it. A CD can store 100–150 songs. In this activity you will use a laser to create a diffraction pattern from a CD. You will then determine the number of lines in 1 cm of a CD.

How can we observe something that is too small to see?

QUESTIONS

- » How does the interference pattern of diffracted light appear?
- » How does the line separation distance affect the maxima spacing in diffraction?
- » How accurately can I determine the CD track spacing with diffraction?

EQUIPMENT

- meter stick
- laser pointer
- wood blocks (2)
- cardboard screen
- protractor
- compact disc
- felt-tip marker
- computer paper (2 pcs.)
- tape

1. Explain in your own words how a straight wavefront can wrap around an object or the edge of an opening in its path.

____

____

____

____

____

2. What experiment conclusively demonstrated the wave nature of light?

____

3. What would be the order number of the third interference band from the central band? How much farther must the waves that follow the longer path from the slits travel compared with the waves traveling the shorter path to form this band?

____

____

Procedure

A Refer to the image shown below as you set up this experiment. Attach the short ends of the two sheets of computer paper. Draw a straight line on the paper using a meter stick. This line will be referred to as the *centerline*. Tape the paper to the table so that the centerline is parallel to the edge of the table.

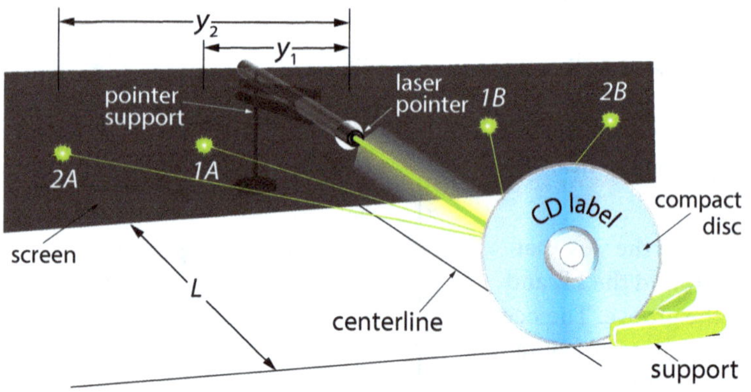

B Determine the wavelength of the light emitted by your laser and record the wavelength in Table 1. This information may be printed on a label, printed on the instrument, or noted in a specifications sheet in the box.

C Draw a centerline on the longest, widest face of one of the wooden blocks. Draw centerlines on the two end faces connected to the line drawn on the long face. Place the laser pointer parallel to and on top of the centerline on the long face. Tape the emitter end of the laser pointer to the block of wood. Set the block on the paper so that the lines marked on the ends of the block line up with the centerline on the paper. Tape the block in this position.

D You will need a hole in the cardboard screen so that the laser pointer will shine through the screen. Position the long side of the cardboard screen so that the laser emitter is centered in the hole and the cardboard rests against the block of wood. Tape the screen to the table and to the block. Use a protractor to verify that the screen is perpendicular to the centerline on the paper and also perpendicular to the table.

E Place the other block of wood on the table. Position the CD with its edge on the table and its label against the block. Tape it in place.

F Position the CD and block assembly so that the centerline on the paper intersects the CD within the data portion of its surface to the right or left of the central hole.

TRIAL 1

Caution: Avoid looking directly into an energized laser, regardless of the size of its beam or the type of laser. All laser emissions are capable of causing eye damage. Be careful of the reflected light too.

G Switch on the laser. You may have to place a piece of tape over the switch to keep the laser energized. Verify that the beam shines parallel to the centerline. Adjust the free end of the laser pointer as necessary to align it; then tape it in place.

H Place the CD and support assembly approximately 15 cm from the screen. Slide the assembly toward the screen until you see a minimum of five bright spots in a line on the screen.

The line of spots should be nearly horizontal, but a small inclination is acceptable. If the pattern is excessively tilted, adjust the laser to shine on the CD at a point where the CD tracks run vertically. Use the protractor to ensure that the CD is perpendicular to the centerline on the paper and adjust the position of the CD so that the bright center spot of the pattern is superimposed exactly on the center of the laser emitter.

I Measure the distance between the surface of the CD and the screen and record this length in Table 1 for Trial 1.

J Mark the centers of the bright spots on the screen with the felt-tip marker, pressing lightly so that the screen does not move during the marking process.

K Label the two inner marks *1A* and *1B* and the two outer marks *2A* and *2B*.

TRIAL 2

L Slide the CD and support assembly away from the screen so that there are only three spots visible on the screen. After realigning the center dot and ensuring that the CD is perpendicular to the centerline, repeat Steps I–K. Label the two visible dots *1A′* and *1B′*.

M Turn off the laser pointer before disassembling the apparatus.

N Measure the separation between maxima for each pair of corresponding Points *A* to *B* (e.g., *1A* to *1B*, *2A* to *2B*, *1A′* and *1B′*). Record these distances in Table 1.

O Determine the spacing of each maximum by dividing the separation between maxima by 2. Enter the results in Table 1.

Analysis

4. Sketch the geometry for the first order maxima of Trial 1. Include the experimental length, the spacing of the maximum, and the diffraction angle.

P Calculate the diffraction angles for each maxima by using the length and the maximum spacing (Step O). Record your answers in Table 2. Show your work for the first order maximum of Trial 1 below.

The track pitch is the spacing between tracks on the CD. This is similar to the distance between etched lines on a diffraction grating.

Name

Q Calculate the track pitch (distance between CD tracks). Record your answers in Table 2. Show your work below for the first order maximum of Trial 1. Also, find the average value of the experimental *d*.

R Find the percent error between your experimentally determined values of track pitch and the actual value of 1600 nm (or 740 nm if you're using a DVD). Record your answers in Table 2. Show your work below for the first order maximum of Trial 1.

5. Discuss the difference(s) between the main spot, the first-order spots, and the second-order spots.

6. Why was there no illumination between the spots?

7. Using this experimental setup, can you obtain a third-order spot, assuming that the light source is 670 nm and the CD track spacing is 1600 nm? Justify your answer.

8. What could you change in order to obtain a third-order or higher interference spot in this experiment?

Conclusion

9. Describe some specific errors that might be introduced if the CD is not parallel to the screen.

Name ______________________

10. If you had used a different CD, do you think that your results would have been different? Explain your answer.

____

____

____

____

11. If the flat cardboard screen were replaced by a curved screen with a given radius of L (the distance between the laser source and the screen), what advantages and disadvantages would this configuration bring to the experiment?

____

____

____

____

____

____

Going Further

12. Why can a CD hold more songs than an LP record?

____

____

TABLE 1 *Diffraction Angle Data*

Trial	Laser Wavelength (nm)	Length (mm)	Order Number	Maxima Separation (mm)	Maximum Spacing (mm)
1			1		
			2		
2			1		

TABLE 2 *Track Spacing*

Trial	Order Number	Diffraction Angle (°)	Track Pitch (nm)	Percent Error (%)
1	1			
	2			
2	1			
		Average:		

LAB

23

Name ______________________

Date ______________________

Absolutely Relative

Investigating the Effects of Relativity

In our everyday experience length and time don't change. A second is a second whether you are standing still, riding in a car, or flying in an airplane. The length of an airplane is the same whether it is stationary on the tarmac or flying at high speed. But are these "facts" always true? Einstein's theories of relativity tell us that these things are *never* true, contrary to our perceptions. The length of an object gets shorter when it is moving. A moving clock runs slow. Einstein's theories demonstrate that these effects are most noticeable when the object is moving really fast, like greater than 10% of the speed of light. Einstein's theories apply even to the everyday objects mentioned above, but the effect is negligible. In this activity you will use simulators to investigate the effects predicted by Einstein's theories.

Why isn't there time dilation and length contraction in everyday motion?

QUESTIONS

- » How can we measure the speed of light?
- » Why might two observers disagree on whether two events are simultaneous?
- » Why do moving clocks run slow?
- » Why must length be contracted in moving objects?

EQUIPMENT

- computer with internet access

Procedure

SPEED OF LIGHT

1. Why is the speed of light so difficult to measure?

____

____

____

____

A Go to the Simulation-Relativity Web Link provided by your teacher. Scroll to *Speed of Light*. Set the simulation to the following initial settings.

Fast Forward/Slow Motion: 1.0

Angular Velocity: 0.2 rad/s

Source Frequency: 4.0 Hz

Speed of Light: 200

B Press the play button.

2. What do you observe about the path of the photons?

____

____

____

3. Predict what you think will happen when you adjust the angular velocity of the rotating mirror. Do the same for adjusting the speed of light.

____

____

____

____

C Adjust the angular velocity to test your prediction. Do the same for your prediction about changing the speed of the photons.

4. Why do you think that the angle changed as it did in each case?

____

____

____

____

____

____

____

D Reset the simulation settings to match those in Step A. When the departing photon stream has become consistent, move the "drag me" detector so that the stream of photons passes directly over the detector.

E Now see what happens to the photon angle when you adjust both the angular velocity and the speed of light simultaneously. Test the following settings for angular velocity and speed of light: 0.1, 100; 0.3, 300; and 0.4, 400.

5. How did the photon angle change when you changed both the angular velocity and the speed of light? Explain.

____

____

Name

SIMULTANEITY

6. What do we mean by simultaneity?

F Using the simultaneity simulator, set Fast Forward/Slow Motion to 1.0 and Velocity to 70%. When the moving figure passes the stationary figure, two lights will flash, one to the left and one to the right of each figure. Each figure will raise his hand when he sees each of the flashes, left hand for the left flash and right hand for the right flash. Start the simulation and watch the stationary figure.

7. Did the stationary figure lift both hands at the same time, indicating that the flashes were simultaneous?

G Pause the simulation as the moving figure approaches the stationary figure. If necessary, slow down the moving figure with the velocity control. With the velocity set to 10%, restart the simulation. Watch whether the two figures agree that the events are simultaneous.

H Adjust Velocity to 20%, click the reset button, and note the time delay between the moving figure raising each of his arms. Repeat for each interval of 10% up to 90%.

8. What happened to the time interval as the velocity increased?

9. Regarding simultaneity, which figure do you agree with? Did this change as the velocity increased? Explain.

10. Predict what will happen when Velocity is set to 100%. Explain.

I Test your prediction.

TIME DILATION

11. Define *time dilation.*

J Opening another browser window, go to the Light Clock Web Link provided by your teacher. Check that Velocity is set to $0c$ and turn on the "Trace." Then start the simulation and notice that the photons follow parallel paths and the clocks run the same amount of time: 8.00 s.

K Press the reset button, then adjust the moving clock velocity to 0.1*c* and turn on the "Trace." The velocity won't change until you release the slider. Then start the simulation.

12. What happened to the paths of the photons and the clocks?

L Repeat Step K using moving clock velocities of 0.2*c*, 0.3*c*, 0.4*c*, 0.5*c*, 0.6*c*, 0.7*c*, 0.8*c*, and 0.9*c*.

13. Predict what will happen when the moving clock velocity is 1*c*. Explain.

M Test your prediction.

14. Describe your observations from Step M.

LENGTH CONTRACTION

15. What is length contraction?

N Go back to the Simulation-Relativity Web Link provided by your teacher. Go to The Necessity of Length Contraction. Set Velocity to 0% and start the simulation. The two photons in each frame are demonstrating that a single flash must travel back and forth in the same amount of time.

O Pause the simulation.

16. What happens to the shape of the contraction when you adjust the velocity slider? What happens when Velocity is adjusted to 100%?

P Start the simulation again and observe the behavior of the photons in each frame of reference as you increase the velocity in increments of 10%.

17. At what point did you notice a change in the motion of the horizontal photon in the moving frame? Explain how the motion was different.

Conclusion

SPEED OF LIGHT

18. The simulator has speeds of light that range from 100 to 400 m/s, but the actual speed of light is 3.00×10^8 m/s. What does your answer to Question 5 imply about the angular velocity of a device that could measure the speed of light?

SIMULTANEITY

19. How well did you predict (Question 10) what would happen to simultaneity when the velocity was set to 100%? Explain.

TIME DILATION

20. Explain your observations from Question 14.

LENGTH CONTRACTION

21. The two photons in each frame of reference had to get back to their starting point at the same time. Why does this imply a necessity for length contraction along the direction of travel?

Going Further

22. As was mentioned in the opener, the effects of simultaneity, time dilation, and length contraction all occur in any moving frame of reference. Consider a 125.32 m long transonic space plane 2816 m/s (Mach 8.500) that takes 1.721 h to fly from London to Sydney, Australia. What is the contracted length of the jet while flying? What is the dilated time interval? Assume that the aircraft flies at a constant speed.

23. According to your calculations, did the length contract and time dilate for this hypersonic flight? Would it be noticeable? Explain.

LAB

24

Name

Date

Fill In the Planck

Determining the Planck Constant

Light-emitting diodes (LEDs) are showing up everywhere, from household lighting to multicolor displays to big screen TVs to ultrabright flashlights. LEDs are constructed of semiconductor materials that emit light when a sufficiently large potential difference—the threshold voltage—is applied across them. The threshold voltage differs for each uniquely colored LED because of quantum physics. Max Planck showed that the energy of a photon of light is related to the light's wavelength by the equation $E = hc/\lambda$, where E is the energy of the photon, λ is the wavelength of the light, c is the speed of light, and h is the Planck constant.

In this lab activity you will put energy into an LED by increasing the voltage to an electric circuit. When sufficient voltage is applied, the diode will release electrons from the semiconductor, emitting photons of light. You will use the threshold voltage and the wavelength of the light emitted by each diode to experimentally determine the Planck constant.

Are there observable things that are related to the Planck constant?

QUESTIONS

» How can I determine the wavelength of light using diffraction?

» How is threshold voltage related to the wavelength of LEDs?

» How can I use the threshold voltage and wavelength of LEDs to determine the Planck constant?

EQUIPMENT

- multimeters (2)
- DC power supply
- computer with spreadsheet software
- diffraction grating
- ring stands (2)
- meter sticks (2)
- test tube clamps (2)
- LEDs, four different colors
- lead wires with alligator clips (5)
- tape

Procedure

A Determine the line spacing of your diffraction grating. This may be recorded on the grating frame, or your teacher will supply this information. Record the spacing in Table 1.

B Using the ring stands, one of the meter sticks, the two test tube clamps, and the diffraction grating, set up the apparatus as shown in the image below. Before installing the diffraction grating, look at a light source in the room through the grating. Orient the grating so that the color spectrum forms to the left and right of the light source. The properly oriented diffraction grating should be aligned with the center of the meter stick, with the diffraction grating and meter stick parallel to each other.

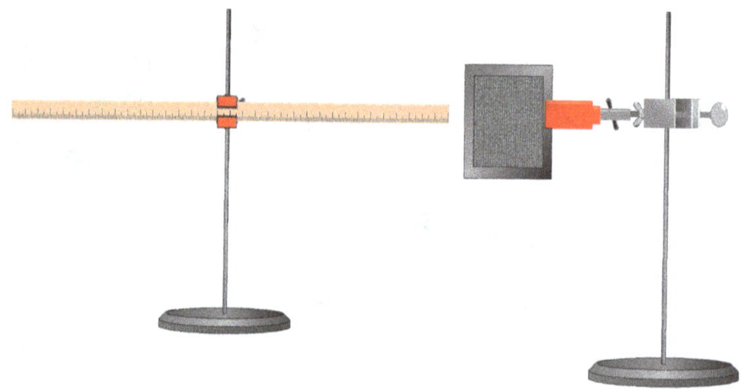

C Set up one of the multimeters as an ammeter (set to approximately 20 mA) and the other as a voltmeter (set to approximately 20 V).

1. Visible light ranges from approximately 450 nm for blue light to 770 nm for red light. According to the energy equations above, which of these colors of light have the highest energy? Explain.

____

____

D Tape one of the LEDs to the center of the meter stick on the support stand. Make sure that you don't cover the light-emitting portion of the LED. Record the color of the LED in Table 1.

E Connect a lead wire from the long wire of the LED to the positive terminal of the power supply. Connect the negative terminal of the power supply to the ammeter. Connect the ammeter to the short wire of the LED. Using the last two lead wires, connect the voltmeter to the long and short wires of the LED. Your circuit should look like the image below. ***Be careful: Connecting the LED in reverse can cause damage to the LED.***

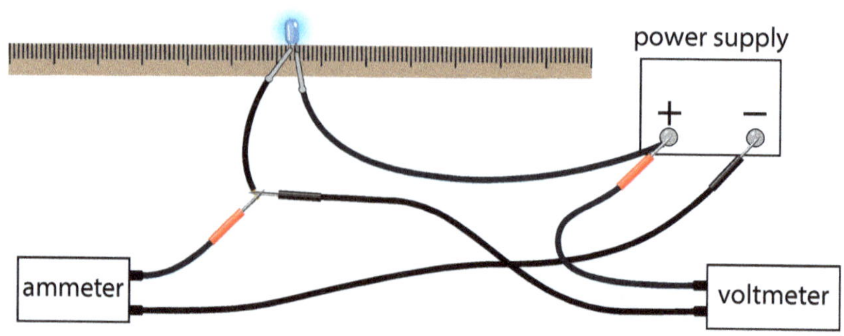

Name ______________________

2. In this lab activity you will not be adjusting energy to the LED; rather you will be adjusting the voltage. Using your answer from Question 1, predict which of the four colors of LED that you expect will need the highest voltage to start emitting light. Explain.

F With the voltage on the power supply set to 0 V, turn on the power supply. Increase the voltage of the power supply until the LED begins to illuminate, then reduce the voltage by approximately 0.25 V. Record the readings from the ammeter and voltmeter in Trial 1 of Table 1.

G Use the fine adjustment knob, if your power supply has one, to increase the voltage by 0.05 V. Record the readings from the ammeter and voltmeter in Trial 2 of Table 1.

H Repeat Step G until you have filled Table 1 for that LED color. ***Do not exceed the voltage limit of the LED.***

I Measure the distance between the diffraction grating and the LED. Record this length in Table 2.

J By looking through the diffraction grating, direct a lab partner to locate the first-order maximum to the left of the LED (see image at right). Record the position of this maximum as Displacement 1 in Table 2.

K Repeat Step J for the first-order maximum to the right of the LED. This is Displacement 2.

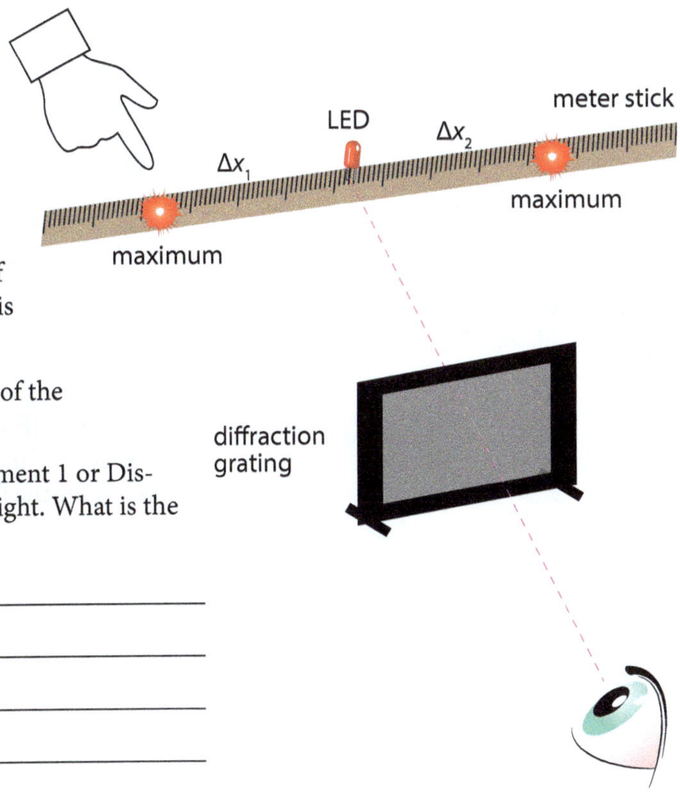

3. We could do this lab activity measuring only Displacement 1 or Displacement 2 to determine the wavelength of the LED light. What is the benefit of using an average of the two values?

L Repeat Steps D–K for the other three LEDs.

Analysis

M Average the two displacements for each LED color and record your answers in Table 2.

4. Use the average displacement and length to determine the average diffraction angle for each LED (see image below). Record your answers in Table 2. Show your work for the first LED below.

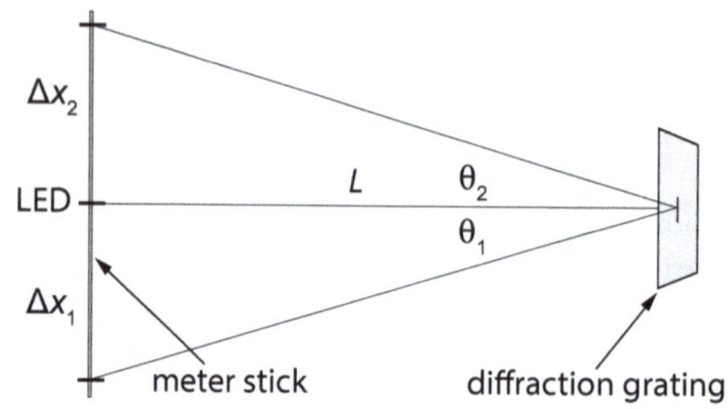

5. Use the diffraction grating equation below to determine the wavelength for each of the LED colors. Record your answers in Table 3. Show your work for the first LED below.

$$\sin \theta = \frac{n\lambda}{d}$$

Name ______________________

N Create a scatterplot of the voltage (y-axis) versus current (x-axis) for each color LED. Create a curve of best fit for the linear (steadily increasing) portion of each graph. Determine the threshold voltage (y-intercept) for each LED and record in Table 3.

6. Threshold voltage is related to wavelength according to the equation

$$h = \frac{eV\lambda}{c},$$

where e is the magnitude of the fundamental charge (1.602×10^{-19} C) and c is the speed of light (3.00×10^{8} m/s). Use this equation to determine the Planck constant for each LED color. Record your answers in Table 3. Show your work for the first LED below.

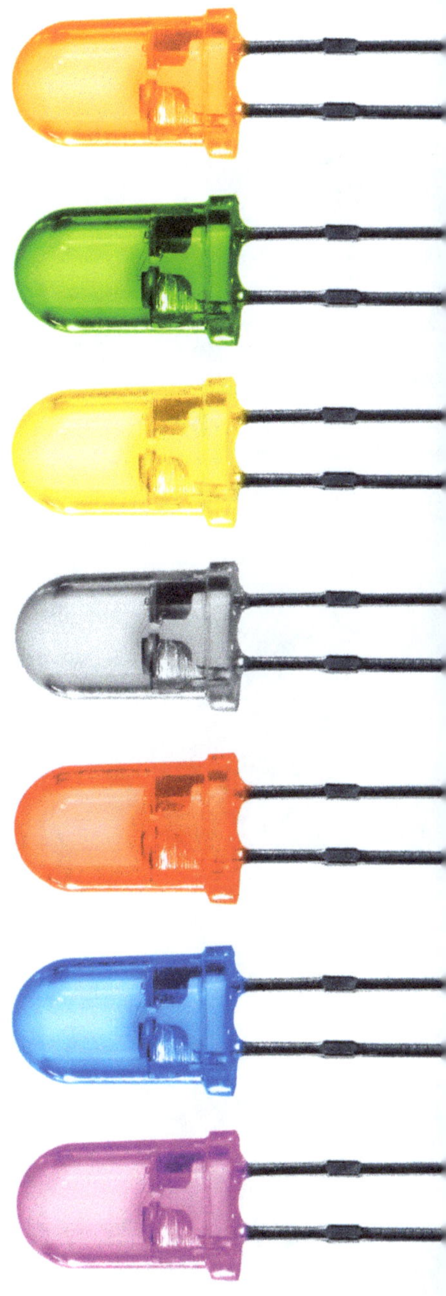

O Calculate the average of the four calculated values for the Planck constant and record this average in Table 3.

7. Calculate the percent error for each of the individual experimental values of the Planck constant as well as the average value. Record your answers in Table 3. Show your work for your first LED below.

Conclusion

8. Were you correct in your prediction in Question 2? Explain.

9. Evaluate the experimental method on the basis of your data.

Going Further

10. Using the voltages and currents obtained in this lab activity as representative values, why are many lighting applications switching to LEDs instead of using incandescent or fluorescent lighting?

Name ______________________

TABLE 1 *Data*

Grating Separation (m)								
LED Color								
Trial	**Current (mA)**	**Voltage (V)**	**Current (mA)**	**Voltage (V)**	**Current (mA)**	**Voltage (V)**	**Current (mA)**	**Voltage (V)**
1								
2								
3								
4								
5								
6								
7								
8								
9								
10								
11								
12								
13								
14								
15								

TABLE 2 *Geometry Data*

LED Color	Length (cm)	Displacement 1 (cm)	Displacement 2 (cm)	Average Displacement (cm)	Average Diffraction Angle (°)

TABLE 3 *Calculations*

LED Color	Wavelength ($\times 10^{-7}$ m)	Threshold Voltage (V)	Planck Constant ($\times 10^{-34}$ J·s)	Percent Error
		Average		

LAB
25A

Name

Date

Halftime Show

Simulation of Radioactive Decay

We often focus on the negative aspects of radioactivity. Radiation can cause some forms of cancer. Many states require radon tests prior to selling a house to minimize people's exposure to that radioactive gas in our homes. But radioactivity can also be beneficial, such as when we use it to produce x-rays and CT scans, enabling us to diagnose injuries and illness. We benefit from available radiation treatments for cancer. Much of our food is irradiated to kill bugs and microbes.

Scientists and archaeologists use radioactivity to estimate the ages of objects that contain radioactive isotopes in a process called radiometric dating. A scientist compares the ratio of the products of radioactive decay (daughter nuclides) to the original nuclides (parent nuclides) in a sample to the assumed ratio when the artifact was made. He can then use the half-life of the isotope to estimate the age of the object. For example, an archaeologist may use carbon-14 dating to determine the date when the particular parchment of the Dead Sea scrolls shown above was made. Other scientists might use potassium-40 dating to determine when a rock was formed. As you can imagine, since we are making interpretations of conclusions about events in the past that are based on data from the present, these analyses are influenced by our worldview. In this lab activity we will investigate radioactive decay and half-life.

How can we predict random events?

QUESTIONS

- What are radioactivity and radiation?
- How can we simulate radioactive decay?
- Will graphs of the radioactive decay of all radioactive isotopes look the same?

EQUIPMENT

- box with cover
- M&M'S® Brand Chocolate Candies (64)

1. Compare the terms *radioactivity* and *radiation*.

2. Summarize alpha decay, beta decay, and gamma decay.

3. Explain why some nuclides are stable while others are not.

4. Describe the variables used in references to the stability and radioactivity of nuclides.

Name ______________________

Procedure

A Count out 64 M&Ms into the box. Do not include any irregularly shaped candies.

B If the candies represent a particular radioactive nuclide, predict how many of the 64 atoms will remain after one half-life. Continue to predict the number remaining after each half-life until none of the radioactive atoms remain. Enter these values in the *Trial 1 Prediction* row of Table 1.

5. Is it possible to know exactly how many radioactive isotope atoms will remain after a certain amount of time elapses? Explain.

____

____

____

____

6. On the basis of your answer to Question 5, why did you make the predictions that you did in Step B? How often will your predictions be correct?

____

____

____

____

C Cover the box. Hold the box horizontally and shake it rapidly up and down several times.

7. What does the process of shaking the box represent about radioactive decay?

____

D Set the box on the table and uncover it. If any candies are resting on top of others, tap the box until they form a single layer on the bottom of the box.

E Remove all candies that show the "m" logo and set them aside. Count the candies remaining in the box and record the number in the *Trial 1* row N_0 column of Table 1.

8. Did the number remaining match your prediction? If not, why not?

F To complete the trial, repeat Steps C–E for each half-life shake until no more candies remain in the box. Record the number of candies remaining for each half-life in the appropriate column of the *Trial 1* row of Table 1.

G Put 64 M&Ms into the box again and repeat Steps C–F for a second trial. Record the results in Table 1.

H Repeat Step G for three additional trials and record your results in Table 1.

Analysis

I Create a single scatterplot of the candies remaining (y-axis) and half-lives (shake number), using the data from Trials 1–5.

9. Did any two of the trials match exactly? Explain.

10. Did any of the trials match your predicted values exactly? Explain.

J Find the average of the five trials for each shake. Record the averages in Table 1.

K Create a scatterplot of the average number of candies remaining (y-axis) and half-lives (shake number).

11. Inspect the graph of the average of the five trials. What is the approximate fraction of candies remaining after each shake?

Name ______________________

12. How many *full* half-lives should be needed to eliminate more than 95% of the radioactive nuclides in a sample? Does your graph support this conclusion?

Conclusion

13. Is a hazardous radioactive material safe after it has been stored for one half-life? Explain.

____

____

14. Your lab group has a single atom nitrogen-13, which has a half-life of 9.97 min. Your lab partner starts a stopwatch and says that the atom will decay in 9.97 min. Is he correct? Explain.

____

____

____

15. How can we predict the decay of radioactive atoms if their decay is random?

____

____

____

____

____

Going Further

16. A scientist in a laboratory is confirming the half-life of carbon-14. He measures the activity of a sample at 1.745 220 Bq. After leaving the sample for 153.2 d, he finds that the activity is 1.745 131 Bq. Using his data, determine the decay constant and half-life of carbon-14.

Name ____________________

17. Carbon-14 decays into nitrogen-14. Scientists can compare the amounts of carbon-14 and nitrogen-14 in an artifact made from organic materials to determine when it died. An archaeologist is conducting carbon-14 testing on one of the Dead Sea scrolls to determine when the parchment was made. He measures 1.352 g of carbon-14 (P_t) and 4.513×10^{-1} g of nitrogen-14 (D_t) in the sample. He assumes that there was 6.62×10^{-2} g of nitrogen-14 in the sample when it was made (D_0). The decay constant for carbon-14 is $1.209 \times 10^{-4}\ y^{-1}$ (λ). Use the model age equation below to determine the age of the scroll.

$$\Delta t = \frac{1}{\lambda}\ln\left(\frac{D_t - D_0}{P_t} + 1\right)$$

TABLE 1 *Radioactive Decay Simulation Data*

Shake	Start	1	2	3	4	5	6	7	8	9	10
Number Remaining	N_0	N_1	N_2	N_3	N_4	N_5	N_6	N_7	N_8	N_9	N_{10}
Trial 1 Prediction											
Trial 1											
Trial 2											
Trial 3											
Trial 4											
Trial 5											
Average											

LAB
25B

Name

Date

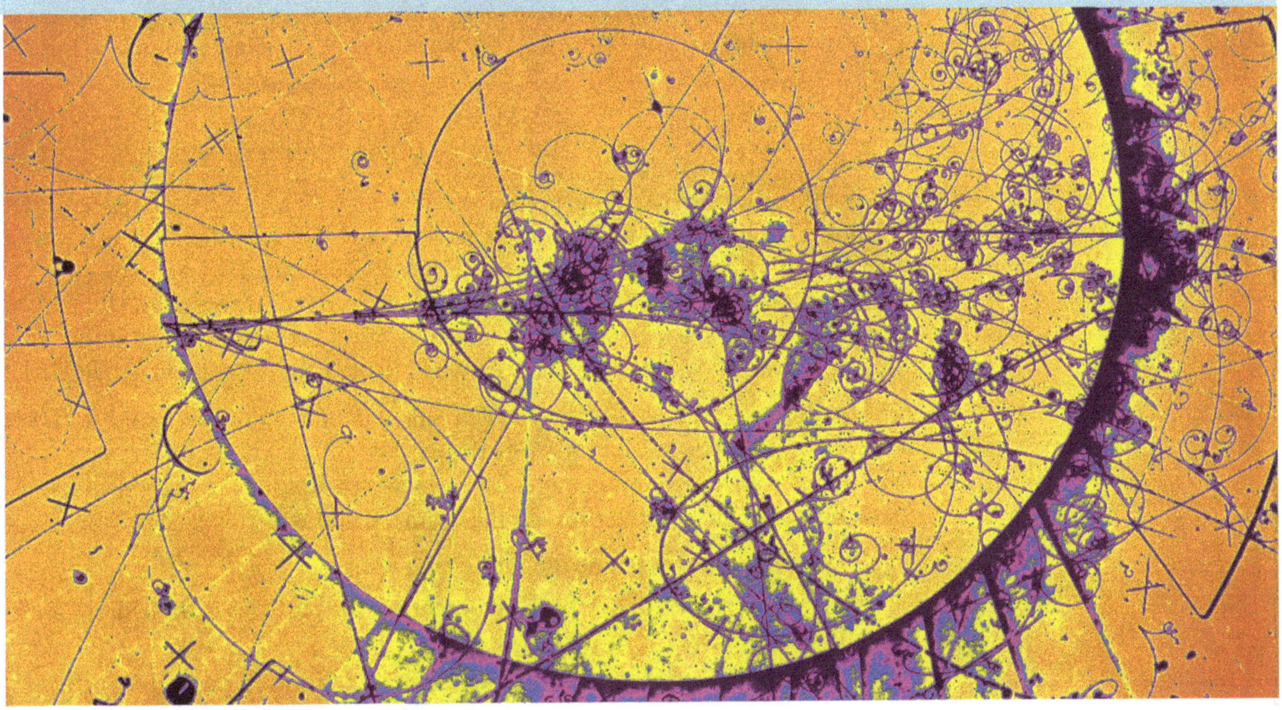

Elementary, My Dear

Investigating Subatomic Particles

Many people become interested in science because they want to know how things work and what they are made of. This is particularly true for physicists in general but especially for particle physicists. What is matter made of? Most of us would say atoms and molecules. Others may mention the subatomic particles—protons, neutrons, and electrons—that make up atoms. And for many years physicists thought they knew all the most basic parts of matter. But then they started making observations that implied that conservation laws were being broken. Either their basic understanding of conservation laws was incorrect or there were subatomic particles that they didn't know about. Scientists have since discovered quite a few additional subatomic particles, some elementary and some composite. Scientists have also hypothesized the existence of even more particles, and today they work to try to observe these particles.

Because of the size of the particles involved, the typical units for mass, momentum, and energy are too large for use with subatomic particles. For example, the mass of a proton is 1.67×10^{-27} kg! Physicists often prefer to work with quantities that are neither very small nor very large. To simplify calculations, scientists express mass, momentum, and energy in units that are more appropriate to the size of the quantities and the kinds of studies being conducted.

How can we study things that we cannot see?

QUESTIONS

- What can we tell about subatomic particles from the data we collect about them?
- Do conservation laws hold true at the subatomic level?

EQUIPMENT

- metric ruler
- protractor

This laboratory activity is based on a lesson developed by Jim Gormley with the assistance of Dr. Alan Jackson, Dr. Didar Qadir, and Dr. Stan Hirschi, professors at Central Michigan University. Used by permission.

TABLE 1 *Relativistic Units*

Variable	Symbol	SI Unit	Subatomic Particle Unit
energy	E	kg·m^2/s^2	MeV
mass	m	kg	MeV/c^2
momentum	**p**	kg·m/s^2	MeV/c

Working with mass and energy at the elementary particle level requires a relationship that should be familiar to all: $E_0 = m_0c^2$, Einstein's famous equation from his theory of special relativity. Physicists use megaelectron volts (MeV) for the very small energies associated with subatomic particles. A particle's rest mass is then simply its rest energy divided by the square of the speed of light. We will use the units shown in Table 1 during this lab activity.

In this "desktop" lab activity you will follow the same procedure that a high-energy physicist uses to identify the particles in a particle accelerator collision. Given a detector chamber image of a particle collision and the magnetic field strength passing through the chamber, you will

1. geometrically determine the radii of the tracks,
2. calculate the magnitude of the momenta of the particles from their track radii,
3. determine the direction of the particle momentum vectors,
4. assess whether momentum was conserved during the collision, and
5. attempt to determine the identity of the particles produced from the collision by identifying their charge and estimating their energies on the basis of hypothetical rest masses.

1. Define *elementary subatomic particle* and *composite subatomic particle.*

____

____

____

____

2. How do scientists simplify working with very small quantities such as the masses, energies, and momenta of elementary nuclear particles?

____

____

____

3. What is the SI equivalent of 1 MeV?

Name ______________________

Procedure

The image at right is a hypothetical picture of a single collision isolated for analysis from a particle accelerator detection chamber image. The projectile is a negative pion (π^-), one of the mesons, which are composite bosons of the hadron group of subatomic particles. The target particle is the nucleus of a liquid hydrogen atom, a proton, which is at rest. The magnetic field strength (**B**) is 3.000 T oriented *into* the page (represented by the ⊗ symbol). The scale size of the bubble chamber is indicated on the photo. To facilitate the analysis of the collision, enlarged "negatives" of the image are provided in Work Areas 1 and 2 on pages 314–15.

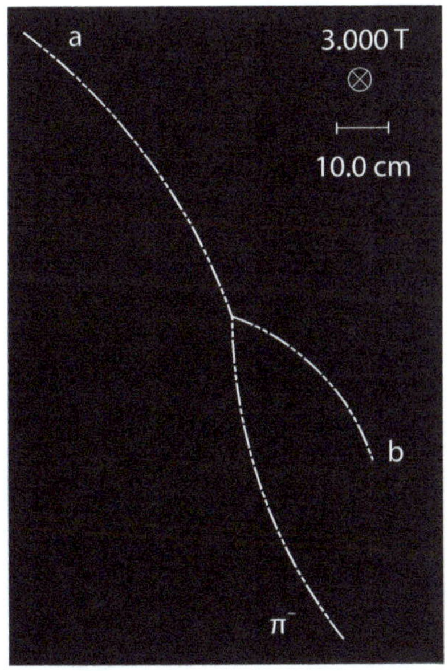

DETERMINE RADII OF PARTICLE TRACKS

The sketch below shows an arc similar to those created by the particles in a bubble chamber. Geometric construction techniques were used to draw a perpendicular bisector of the arc. The bisector, the chord between the two endpoints, and some additional segments create a triangle. From the triangle we can derive an equation for the radius of the arc.

A Use the sketch to derive a formula for the radius of an arc if we can measure a and b but h is unknown. Show your work below.

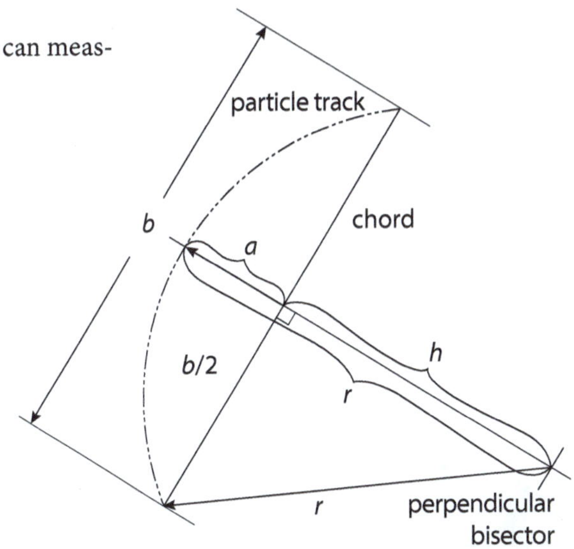

The triangle formed by completing a geometric construction to find the radius of an arc

B Using the arcs formed by the three particles shown on Work Area 1, measure the length of the chord (b) and the distance between the chord and the arc (a). Then use the formula derived in Step A to determine the radius of each of the particle paths. Record these radii in Table 2. Show your work for the pion below.

4. Why are there no paths shown for the proton?

____

____

____

C A scale line is provided in the diagram in the Work Area. Use this scale line to determine the scale factor for the diagram and record in Table 2.

D Use the scale factor to determine the actual radius of each of the particle paths in the bubble chamber and record in Table 2.

The concept of momentum and the law of momentum conservation should be valid for microscopic objects as well as for macroscopic ones. You should be able to apply these ideas to molecules, to atoms, and even to subatomic particles.

Name ____________________

CONSERVATION OF MOMENTUM

Before continuing further, study the tracks in the hydrogen bubble chamber shown in the image at right. The curves are the tracks of particles. Points where the tracks split are either collisions between the projectiles and protons (the nuclei of liquid hydrogen) or the decay of particles into other particles. If you lay a straightedge along the tracks, you will notice that not one is straight. They form these arcs because of an external magnetic or electrical force acting on the charged particles perpendicular to the particles' motion. Recall from studying circular motion that the formula describing circular motion is

$$F_c = m a_c = \frac{m v_t^2}{r}, \qquad (1)$$

where F_c is the magnitude of the centripetal force, m is the mass of the object, a_c is the centripetal acceleration, v_t is the object's tangential velocity, and r is the radius of the circular path.

In the case of the bubble chamber image for this lab activity, the external force exerted on the particles is a magnetic force produced by a magnetic field that is generated around the bubble chamber. Recall from your study of magnetic fields that the magnitude of the magnetic force can be calculated by using the formula

$$F_m = q v_t B \sin\theta,$$

where q is the particle's charge in multiples of the fundamental charge e, v_t is the particle's velocity, B is the magnetic flux density—magnitude of the magnetic field—in teslas, and θ is the angle between the velocity and magnetic field vectors. Since the magnetic field is perpendicular to the particle velocity vector, the magnetic force equation simplifies to

$$F_m = q v_t B. \qquad (2)$$

Since the circular motion of the particles is due to the application of the external magnetic field, we can set Equations 1 and 2 equal to each other. We can then derive a formula for the momentum of the particles in the bubble chamber.

$$F_c = F_m$$

$$\frac{m v_t^2}{r} = q v_t B$$

$$\frac{m v_t^{\cancel{2}}}{\cancel{r}} \cdot \frac{\cancel{r}}{\cancel{v_t}} = \frac{q \cancel{v_t} B r}{\cancel{v_t}}$$

$$m v_t = q r B$$

Since $mv_t = p$, we have a formula for the magnitude of the momentum of particles in the bubble chamber.

$$p = qrB \tag{3}$$

For particles with a charge of 1*e*, such as protons and electrons, Equation 3 may be simplified to

$$p = 3rB, \tag{4}$$

where the particle momentum, p, is measured in MeV/c, r is measured in centimeters, and B is measured in teslas. The coefficient in Equation 4 is simply a conversion factor to convert cm·T to MeV/c. The conversion factor is shown below.

$$\frac{3\ \text{MeV}/c}{1\ \text{cm}\cdot\text{T}}$$

E Assuming that all three particles have a charge of 1*e*, calculate the magnitude of momentum for each particle. Enter your results in Table 2. Show your work for the pion below.

F Draw coordinate axes positioned at the point of collision on the Work Area 2. The axes should be perpendicular, but their orientation relative to the particle tracks is not critical.

G Recall that the linear velocity of an object undergoing circular motion is tangential to the curve. Measure the angle for the direction of the momentum for each particle at the instant of collision and record the angle measures for each particle in Table 2.

Name ____________________

H Determine the x- and y-components of the magnitude of each particle's momentum. Be sure to observe correct sign conventions according to your coordinate system. Record all results in Table 2. Show your work for Particle a below.

I Determine the total momentum vector components before and after the collision by summing the components for each particle and record your results in Table 2.

J Use the total momentum vector components before and after the collision to determine the total momentum before and after the collision. Record your results in Table 2. Show your work below.

CONSERVATION OF ENERGY

In classical mechanics the total mechanical energy of an object is determined by the sum of its kinetic and potential energies: $E = K + U$. And this relationship is true within 1% for speeds less than 10% of the speed of light. However, the particles that produce the hydrogen bubble chamber tracks are traveling near the speed of light. The momentum of a particle under these conditions is determined by the equation $p = \gamma m_0 v$, where γ is the Lorentz factor that accounts for the relativistic speeds. For this reason, the relationship between total energy and momentum needs to be adjusted for relativistic speeds. Einstein showed that the relativistic energy-momentum formula is

$$E^2 = (pc)^2 + (m_0c^2)^2, \tag{5}$$

where pc represents the particle's energy due to momentum, and m_0c^2 is its rest energy (in MeV). You should note that a particle at rest has no momentum, so Equation (5) in this case reduces to $E = m_0c^2$. Just as momentum is conserved in a collision, total energy must also be conserved in these elastic collisions.

K Using the momenta determined in Step E and the rest energies from Appendix J, write the energy-momentum equation (Equation 5) for the two particles (pion and proton) involved in the collision in Table 3. Then calculate the total energy for each of those particles and record your results in Table 3. Show your work for the pion below.

L Calculate the total energy before the collision in Table 3.

M Using the momenta determined in Step E, write the energy-momentum equation (Equation 5) for the two particles (a and b) resulting from the collision in Table 3. Since we don't know the rest energy for these unknown particles, you will need to represent the rest energy with variables for now.

Name ______________________

Since you do not know the rest energies (masses) of Particles *a* and *b*, which is the case in most HEP experiments, you need to make a guess at the rest energies so that you can complete the calculations for total energies. If you select the correct rest energies for the unknown particles, total energy should be conserved, and you can be relatively sure that those are the rest energies of the unknowns. From the law of conservation of energy, we can conclude that

$$E_{\pi^-} + E_p = E_a + E_b. \tag{7}$$

Before calculating the total energies of candidate unknown particles, you should try to limit the potential candidates by considering their charge properties. Note from the image in the Work Area that *a* curves to the left in the magnetic field and *b* curves to the right.

5. What can we conclude about the charge on Particles *a* and *b* since they curve to the left and to the right respectively? Explain.

____

____

____

____

____

____

N Determine the charges of Particles *a* and *b* by noting the direction of curvature of their tracks in the image in the Work Area. Write the charge for each particle in Table 3.

O Refer to the table of subatomic particles in Appendix J. Enter the symbol, including its charge, and its rest energy for both Particle *a* and Particle *b* in Table 4.

P Using the momenta determined in Step E and the rest energies that you obtained from Appendix J, calculate the total energy for each of those particles and record your results in Table 4.

Q Select a positive particle and a negative particle from the candidates in Table 4 such that the sum of their total energies equals the total energy before the collision.

R Write the identities and total energies of Particles *a* and *b* in Table 3. Record their total energies and the sum of their energies in Table 3.

Analysis

6. State your conclusions regarding the identities of the particles resulting from the collision.

____

____

____

7. If you knew the speed of the particles, how could you determine the length of time represented by each of the tracks?

____

____

____

Conclusion

8. Considering the conservation principles that you are studying, do you think that your answer to Question 6 is reasonable?

____

____

____

____

____

Going Further

9. If the track of a particle in a detector chamber is an inward spiral, what can you conclude about the momentum of the particle during the time period in which the track was created? Explain your answer.

____

____

____

TABLE 2 *Momentum Analysis*

Scale Factor:

Particle	Arc Radius (cm)	Particle Radius (cm)	Particle Momentum (MeV/c)	Momentum Reference Angle (°)	Momentum *x*-component (MeV/*c*)	Momentum *y*-component (MeV/*c*)
Before the Collision						
$\pi-$						
p						
Total Momentum, Components (MeV/*c*)						
Total Momentum (MeV/*c*)						
After the Collision						
a						
b						
Total Momentum, Components (MeV/*c*)						
Total Momentum (MeV/*c*)						

Name ______________________

TABLE 3 *Energy Analysis*

Particle	Symbol	Charge	Energy-Momentum Equation	Energy (MeV)
Before the Collision				
Pion	π^-	−		
Proton	p	+		
Total Energy				
Before the Collision				
a				
b				
Total Energy				

TABLE 4 *Candidate Particle Total Energies*

Particle *a*			Particle *b*		
Candidate	Rest Energy (MeV)	Total Energy (MeV)	Candidate	Rest Energy (MeV)	Total Energy (MeV)

Work Area 1

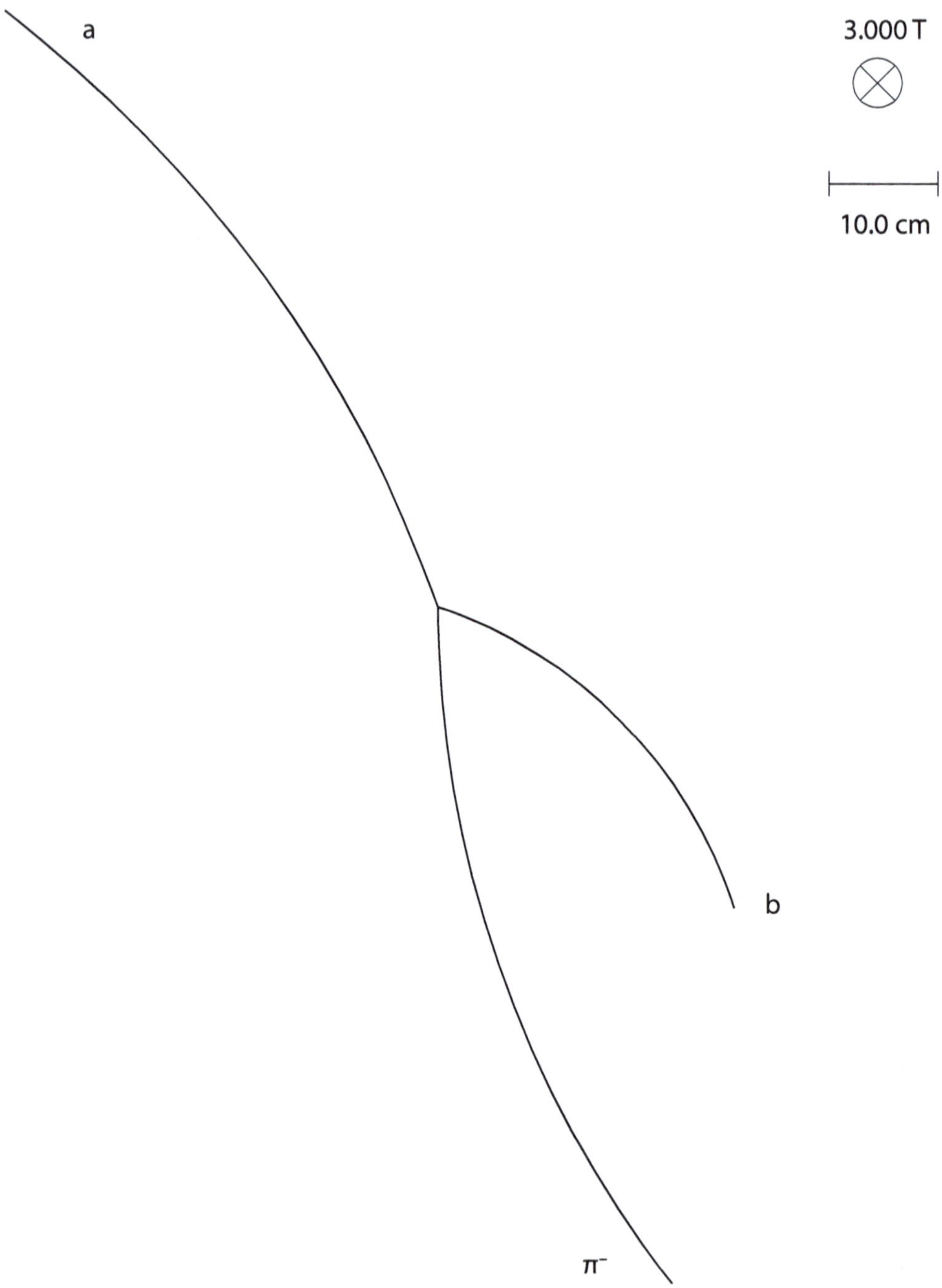

Name ______________________

Work Area 2

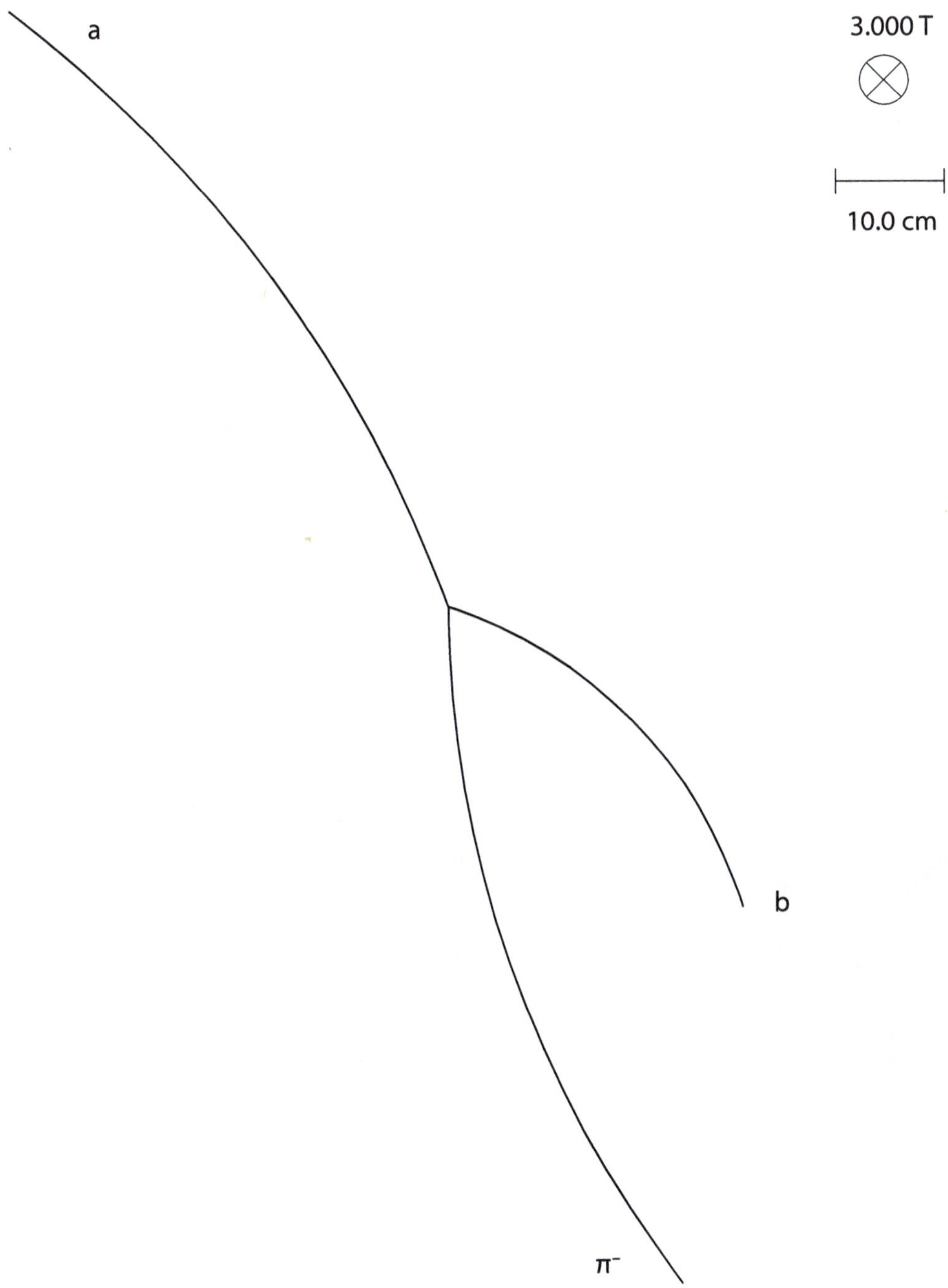

APPENDIX A
Laboratory and First-Aid Rules

LABORATORY RULES

1. Never perform an unauthorized experiment or change any assigned experiment without your teacher's permission.
2. Avoid playful, distracting, or boisterous behavior.
3. Never work alone. Students must not conduct lab activities without supervision.
4. Work at your own lab station.
5. Always work in a well-ventilated area. Use the fume hood when working with toxic vapors. Never put your head in the fume hood.
6. Always wear safety goggles when working with chemicals, glassware, projectiles, and other materials or objects that are potentially hazardous to the eyes.
7. Wear protective clothing and gloves when working with corrosive or staining chemicals.
8. While working in the laboratory, tie back long hair and avoid wearing loose clothing such as scarves or ties.
9. Never taste any chemical, eat or drink out of laboratory glassware, or eat or drink in the laboratory.
10. Always use the appropriate instruments for cutting and handle them carefully. Always cut away from yourself.
11. Thoroughly wash your hands with soap after handling any chemicals.
12. To smell a substance, gently fan its vapor toward you.
13. Never leave a flame or heat source unattended. Keep combustible materials away from heat sources.
14. When diluting acid solutions, always add the acid to water slowly. ***Never add water to an acid!***
15. When heating a test tube, point the open end away from you and others. ***Never heat a closed or stoppered container!***
16. Dispose of waste as instructed by your teacher.
17. Do not return unused chemicals to a container. Dispose of them properly.
18. Notify the teacher of any injuries, spills, or breakages.
19. Know the locations of the fire extinguisher, safety shower, eyewash station, fire blanket, first-aid kit, and Safety Data Sheets (SDS).

APPENDIX A
Laboratory and First-Aid Rules

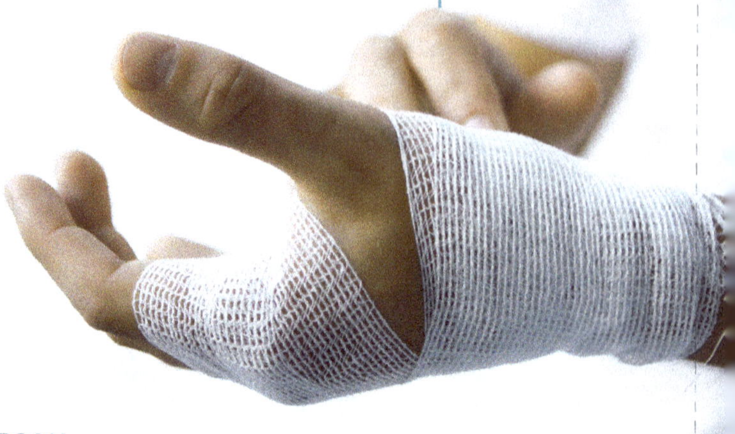

BASIC FIRST AID

Burns

Flush the area with cold water for several minutes. Do not apply ice.

Chemical Spills

Notify your teacher of all chemical spills.

ON A LABORATORY DESK

1. If the material is not particularly volatile, toxic, or flammable, your teacher may have you clean the spill. For liquids, use an absorbent material that will soak up the chemical. For solids, use the designated dustpan and brush. Dispose of chemicals and cleaning materials properly. Then clean the area with soap and water.
2. If the material is volatile, toxic, or flammable, and not a large spill, ask your teacher for help. If it is a large spill, you may need to evacuate the laboratory.
3. If a highly reactive material, such as hydrochloric acid, is spilled, your teacher will clean it up.

ON A PERSON

1. If the spill covers a large area, begin rinsing in the chemical shower, then remove all contaminated clothing and remain under the safety shower. Flood the affected body area for fifteen minutes. Obtain medical help immediately.
2. If the spill covers a small area, immediately flush the affected area with cold water for several minutes. Then clean the area with soap and water. Get medical attention.
3. If the chemical splashes in your eyes, immediately wash them in the nearest eyewash fountain for at least 15–20 minutes. Get medical attention.
4. If the spill is an acid, rinse the area with sodium bicarbonate or sodium carbonate solution; if it is a base, use citric or ascorbic acid solution.

Fire

a. For any fire other than a contained fire, do *not* attempt to put it out on your own. *Understand that some fires can't be put out with water.*

b. Smother a small fire in a container by covering it.

c. If a person's clothes are on fire, remember to stop, drop, and roll—roll the person on the floor and use a fire blanket to extinguish the flames. The safety shower may also be used. ***Do not use a fire extinguisher!***

Swallowed Chemicals

Determine the specific substance ingested and follow the instructions on the SDS. Contact the Poison Control Center in your area immediately.

Cuts

If the wound is superficial, clean it with a disinfectant, apply triple antibiotic ointment, and cover with a bandage. If it is a deep cut, seek immediate medical attention.

APPENDIX B
Laboratory Techniques

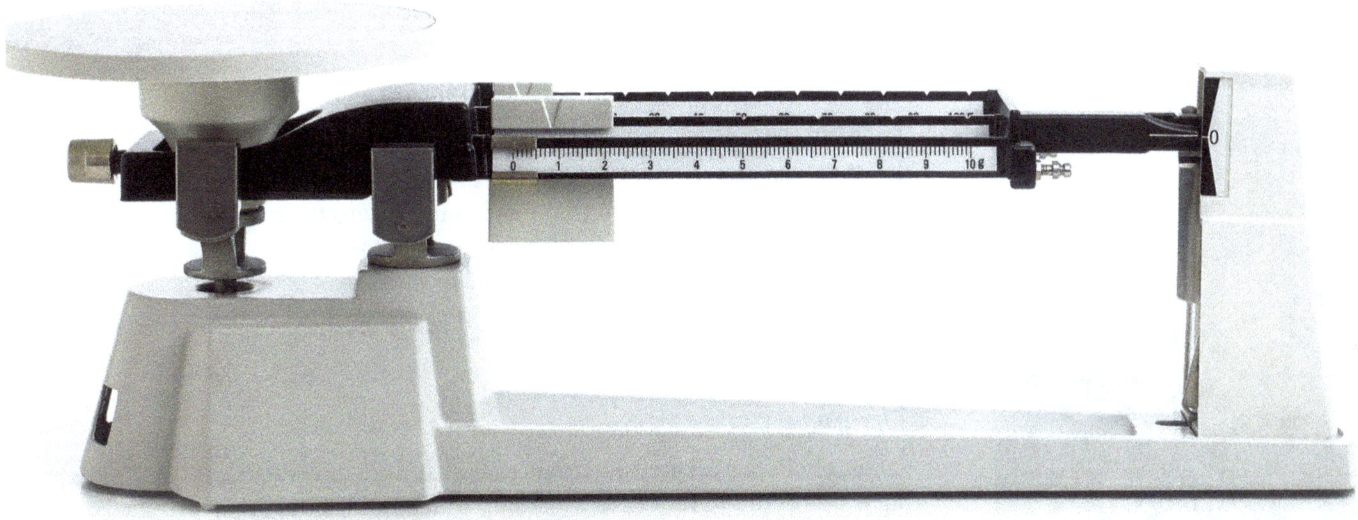

Measuring Mass

USING A MECHANICAL BALANCE

The mass of a substance can be determined in the laboratory with the use of a mechanical balance. Several kinds of mechanical balances are common, but all of them operate on the same principles. To use a mechanical balance properly, follow the steps given below.

1. Place the balance on a smooth, level surface.
2. Keep the balance pan(s) clean and dry.
3. Check the rest point of the empty balance. To do this, remove all masses from the pans and slide all movable masses to their zero positions. If the balance beam swings back and forth, note the central point of the swing. You do not have to wait until the beam stops swinging completely. If the central point is not aligned with the marked zero point, ask your teacher to adjust the balance. *Do not adjust the balance yourself.*
4. Place the substance to be massed on the pan and adjust the sliding masses. Move the largest masses first, and then make final adjustments with the smaller masses. If the beam has detents for the sliding masses, make sure that the masses rest in the detent. The sum of all the readings is the mass of the object. The mass of the sample in the image shown at right would be read as 101.43 g.

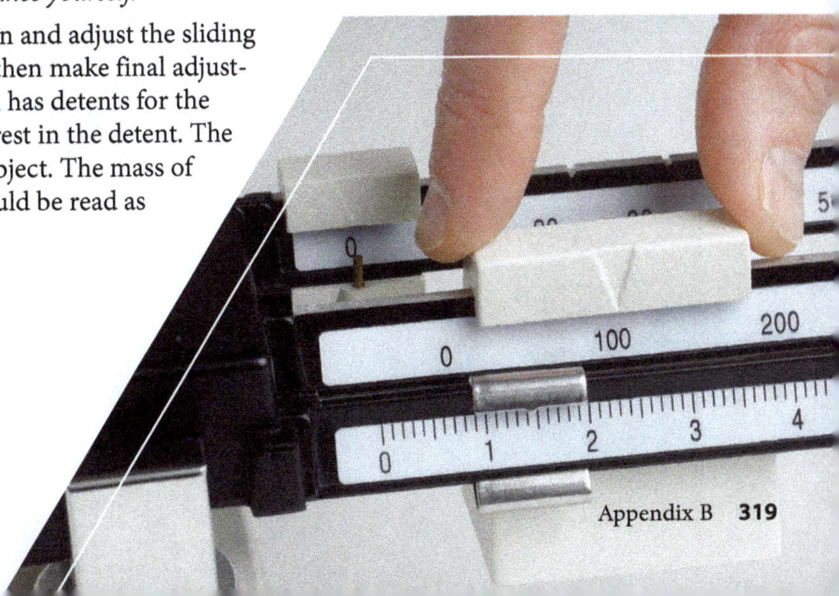

APPENDIX B
Laboratory Techniques

USING AN ELECTRONIC BALANCE

Electronic balances are generally faster and easier to use than their mechanical counterparts. To use an electronic balance properly, follow the instructions given below.

1. Place the balance on a smooth, level surface.
2. Keep the balance pan(s) clean and dry.
3. Turn the balance on. Press the **Tare** button to zero the balance. If you will mass the object in a container, you can zero the balance with the container on the balance pan. Place the container on the balance and press the **Tare** button to zero the balance.
4. Place the object in the container or directly on the pan. Read the measurement from the balance.

Timing Events

Often in physics we need to measure the time interval of an event. Timing is procedurally very easy. Most stop watches, including those on smartphones, are set up with start, stop, and reset buttons. Some stopwatches also have a lap button that allows an intermediate time to remain visible while continuing to time the entire event. The real challenge with measuring time intervals is to accurately start the stopwatch at the beginning of the event and stop it at the end of the event. This difficulty is eliminated by using an automated trigger for starting and stopping the stopwatch. If you are going to time events by hand, work with your lab group members to determine the best method for obtaining accurate time interval measurements.

Using a Thermometer

When taking a reading, position the thermometer bulb just above the bottom of the container. If the bulb touches the container, your readings may be inaccurate. Always hold a thermometer when it is in a container. *Never leave a thermometer standing upright in a container—the container is likely to get knocked over.* If a thermometer breaks, alert your teacher and do not touch the inner contents. Some thermometers contain mercury, which may look fascinating but is toxic and can be absorbed through the skin.

Using Electrical Test Instruments

A variety of meters is available to measure properties of electric circuits, such as resistance, voltage, and current. Some may be designed to measure just one quantity, or they may measure several different quantities, as *multimeters* do. Electric meters are precision instruments, so they must be handled with reasonable care. The most common causes of damage are dropping a meter or incorrectly connecting it to a circuit. If you are not sure about how you have connected your circuit, check the diagrams in the lab activity that you are working on; you can also check with your teacher. An *analog meter*—a meter that provides continuous indication throughout its range—requires special care. Analog meters have a needle attached to a mechanism known as a *meter movement*. They are typically designed for use in either AC or DC circuits and are not usually interchangeable. Many meters of this kind are designed for a specific range of values. The meter movement is very delicate and can be damaged easily. *Digital meters* measure an electrical quantity by sampling the quantity of interest, converting the samples into a numeric value, and displaying the number on a display.

Though many meters include a fuse or circuit breaker, a safety feature that protects the meter, always use good procedures when working with electrical devices to avoid damage. First, whenever possible, insert a switch in the circuit that will allow you to quickly open the circuit if the current is too high for the instrument. Second, if the meter has manual range selection, choose the highest scale to make the initial reading and then adjust the range scale downward until you obtain readings that fall in the middle of the scale (when using an analog meter) or provide the greatest possible number of significant figures (when using a digital meter). Third, for analog meters, ensure that the connection polarity is correct; this will prevent the pointer from being slammed backward against the meter stop, possibly damaging the meter movement.

MULTIMETERS

Multimeters can be either analog or digital and can measure several electrical quantities. Many multimeters have sophisticated "autoranging" circuits that safely select a range appropriate to the magnitude of the quantity being measured. Other meters usually require that the range be selected manually. Multimeters are convenient because you can easily switch from one function to another by plugging one of the test leads into a different terminal on the meter or by using a selector switch. Basic models include voltmeter, ammeter, and ohmmeter functions, and most can measure both AC and DC. Analog multimeters will have a variety of scales on display for the different quanties that they can measure. Digital multimeters have a single liquid crystal display that automatically changes with the meter mode. Many multimeters have other functions, such as signal level, electrical continuity, transistor or diode function, and even temperature with the aid of a thermocouple probe.

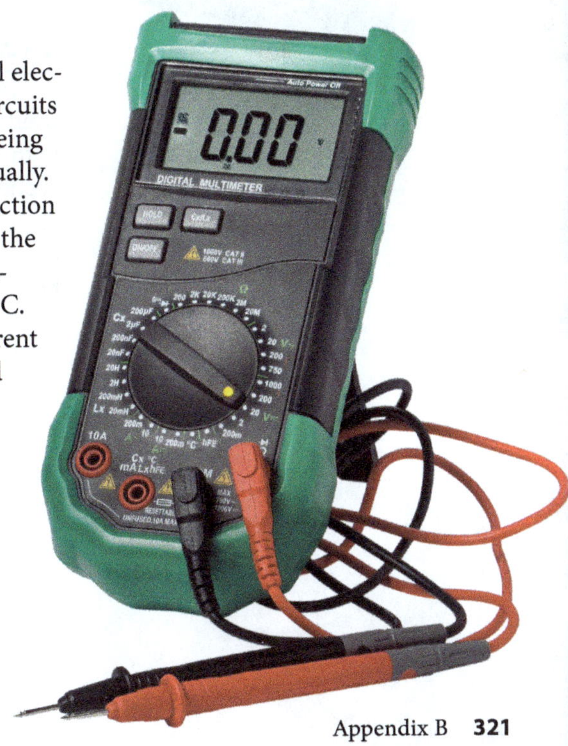

APPENDIX B
Laboratory Techniques

VOLTMETERS

A voltmeter is used to determine the potential difference between two points in a circuit. It is connected in parallel with the component being measured since voltage of parallel components are equal. A voltmeter's internal resistance is very high so that it will not alter the current in the circuit being measured. The terminals of a voltmeter are marked with plus and minus signs. The red test probe is usually connected to the positive terminal and the black probe to the negative terminal. To avoid negative voltage readings on digital meters or backward pointer movement with analog meters, always connect the positive terminal of the voltmeter to the end of the component closest to the positive side of the voltage source (via the circuit) and the negative terminal to the negative side of the component.

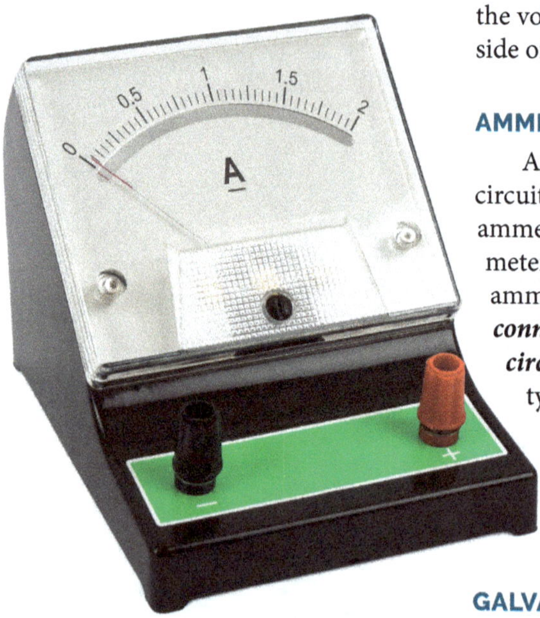

AMMETERS

Ammeters are used to determine the amount of current flowing in a circuit. Since the current is constant for series components, we connect ammeters in series with the section of the circuit being measured. Ammeters are designed with low internal resistance because we don't want the ammeter to cause a drop in voltage. ***Warning: Due to its low resistance, connecting an ammeter in parallel can easily cause an overcurrent in the circuit.*** Note that standalone ammeters for AC and DC are different. One type cannot produce an accurate indication in the other kind of circuit. Multimeters often have different ammeter selections for AC and DC current. As with the voltmeter, ensure that the positive and negative terminals on the instrument are connected in the circuit with respect to the corresponding terminals of the current source.

GALVANOMETERS

A galvanometer is essentially an extremely sensitive ammeter that is used to measure very small currents (typically in the microamp range). As with regular ammeters, a galvanometer must be connected in series with the section of the circuit being measured. It can indicate the presence of a current, its direction, and its relative magnitude. Some galvanometers may have a low-resistance wire across the terminals to protect the instrument from being exposed to currents that are too high. Galvanometers typically have a sensitive protective fuse. Despite these features, however, galvanometers are easily damaged.

OHMMETERS

Ohmmeters measure the resistance of an electrical component. An ohmmeter contains a battery and a very precise voltage regulator circuit to generate a current through the component in proportion to its resistance. For this reason, the component must not be powered and should be disconnected from the circuit before measuring its resistance. ***Warning: If an ohmmeter is connected to an energized circuit, it can be damaged.***

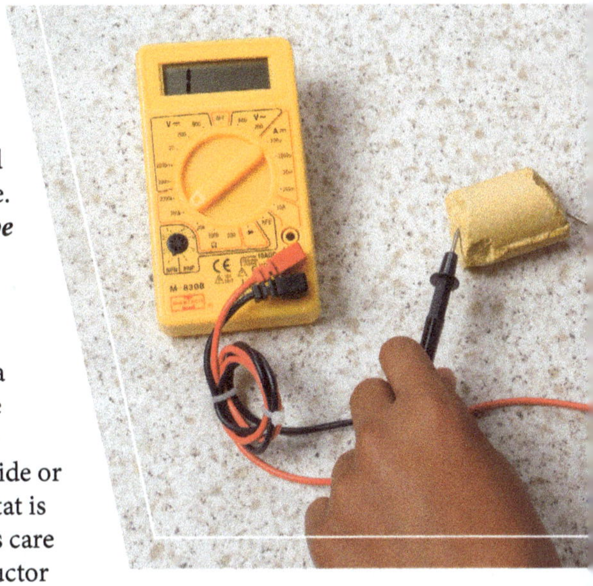

RHEOSTATS

A rheostat (also called a variable resistor) is a device inserted into a circuit that allows for varying the resistance and the current flow in the circuit. High-resistance wire in an instrument is not greatly affected by temperature changes, making it useful for laboratory experiments. A slide or a rotary wiper allows the resistance to be changed as required. A rheostat is not technically a meter, but it is a delicate electrical device that requires care in use. Sharp impacts on the wiper surface can short or open the conductor windings in the device, destroying its function.

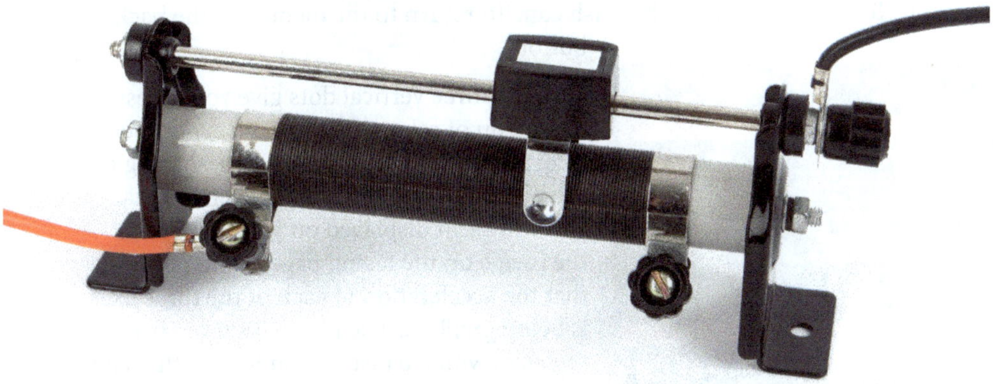

APPENDIX C

Using the Phyphox® App

In addition to providing a ready source of communication and entertainment, our smartphones are also very capable physics laboratories. Several apps allow you to use the sensors in your phone to collect data from physics experments. A number of lab activites in this lab manual use the Phyphox® app. This appendix will provide you some support for using the app. If you choose to use a different physics app, most of them operate in a manner similar to how the Phyphox® app works.

Once you have downloaded the Phyphox® app, begin by opening the app. You will see a page, like the one shown below, that lists all the different experiments available in the app. Select the experiment that you want to perform. Keep in mind that most of the experiments will show the collected data on graphs. The image on the facing page, left, shows the Acceleration (without **g**) experiment. Note that this experiment will generate three acceleration-time graphs, one for each axis. Across the top of the graph areas there will often be different options for the data that will be displayed. Start data collection by pressing the start button located in the orange menu bar at the top of the app. To discard data, use the trash can. To return to the menu, use the back button.

Opening page

The three vertical dots give you access to a menu of other options (e.g., experiment information, exporting your data, sharing a screenshot).Once you start collecting data, you will see it displayed on your screen. In the image on the facing page, right, notice that the acceleration in each of the three axes is being graphed. The play button has been replaced with a pause button to stop the data collection. Once you have your data, you can easily send it to a computer using the export option. You can send the data as an Excel spreadsheet file or a csv (comma separated values), which can be put into other spreadsheet programs on a computer.

If you keep the app updated, new experiments may become available as developers create them. You can access additional updated information on the Phyphox® page through the app.

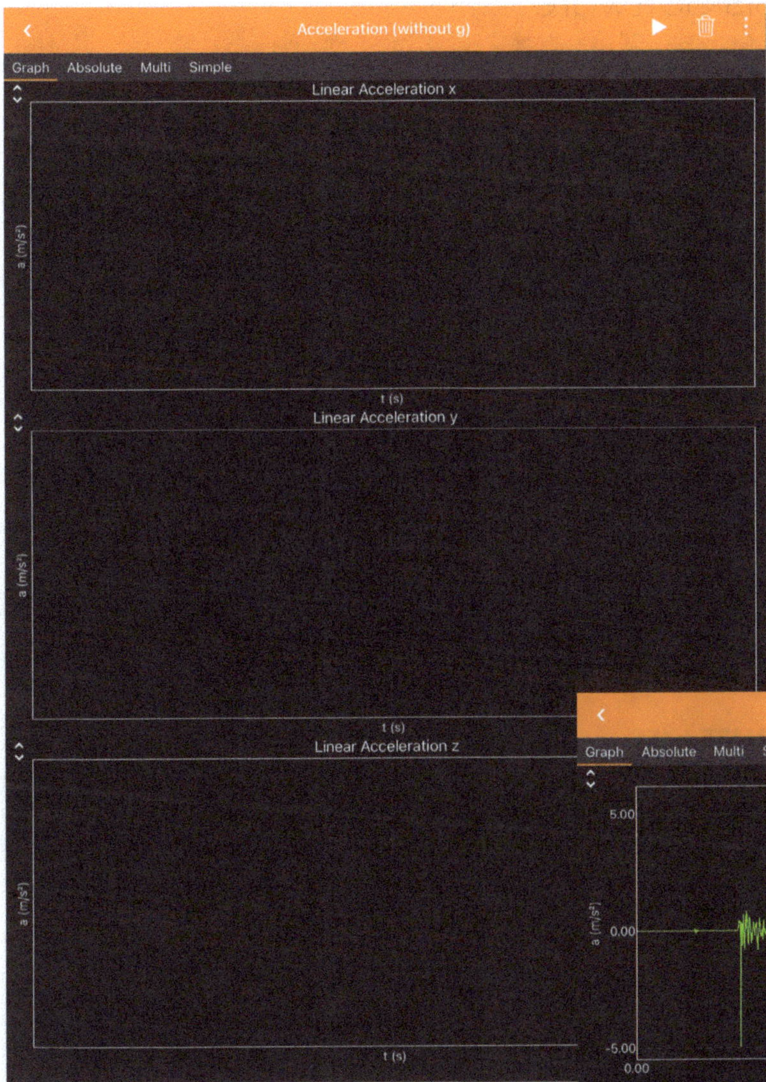

The Acceleration (without **g**) experiment screen

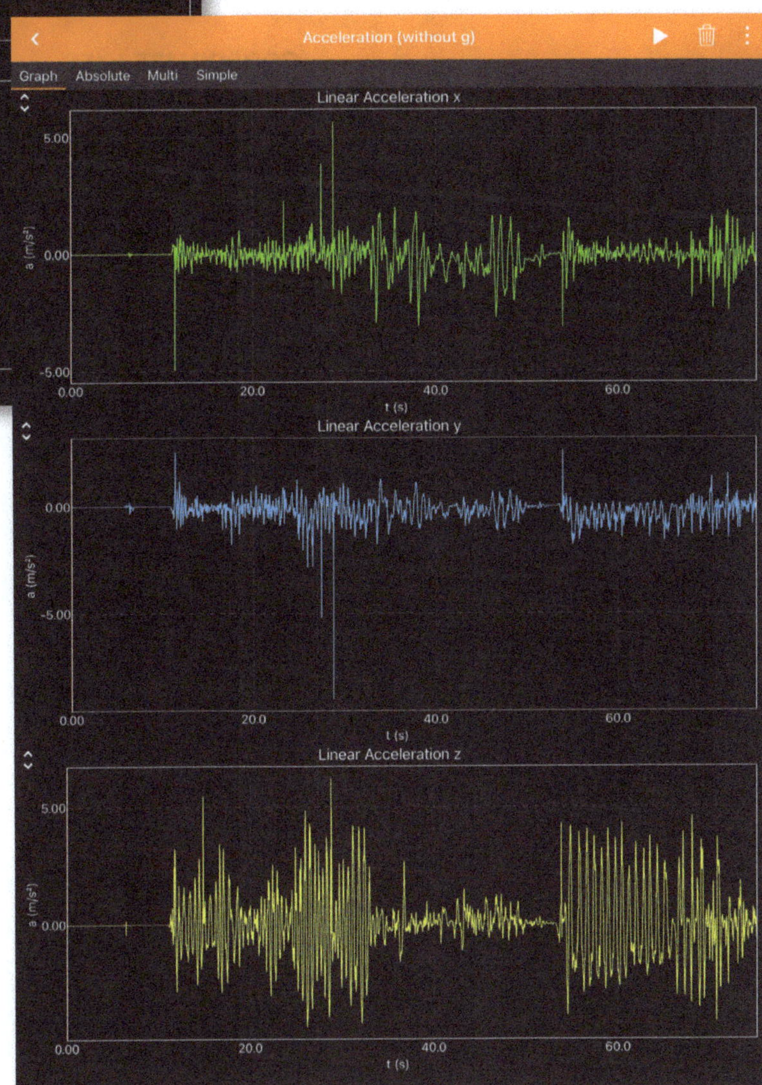

Three-axis acceleration data shown on graphs

APPENDIX D
Tracker Video Analysis Software

There are many programs for analyzing videos. Tracker is an open source program that is readily available. The following covers the basics of using Tracker. (On the Tracker video analysis website you will find many tutorial videos and help resources for any other questions that you have.) While a high-end video camera will produce excellent videos for analysis, current smartphone cameras can take videos of sufficient resolution to be used for video analysis. Once you have shot your video, send it to your computer. The rest of your analysis can be done in Tracker or with a spreadsheet program.

Importing Videos

Before you can analyze a video, you need to import it into the program. On the **Video** menu select **Import**. Then navigate to the location of your video file.

Setting Up the Analysis

Once you have loaded a video, set up for the analysis. You will need to select your axes, set your scale, and set your start time.

COORDINATE AXES

Notice the toolbar at the top of the program (see below). If you click on the **Show or Hide the Coordinate Axes** button, coordinate axes will be superimposed over the video. You can drag the axes to wherever you want the origin. You can even drag one of the axes to slant the axes for inclined plane problems.

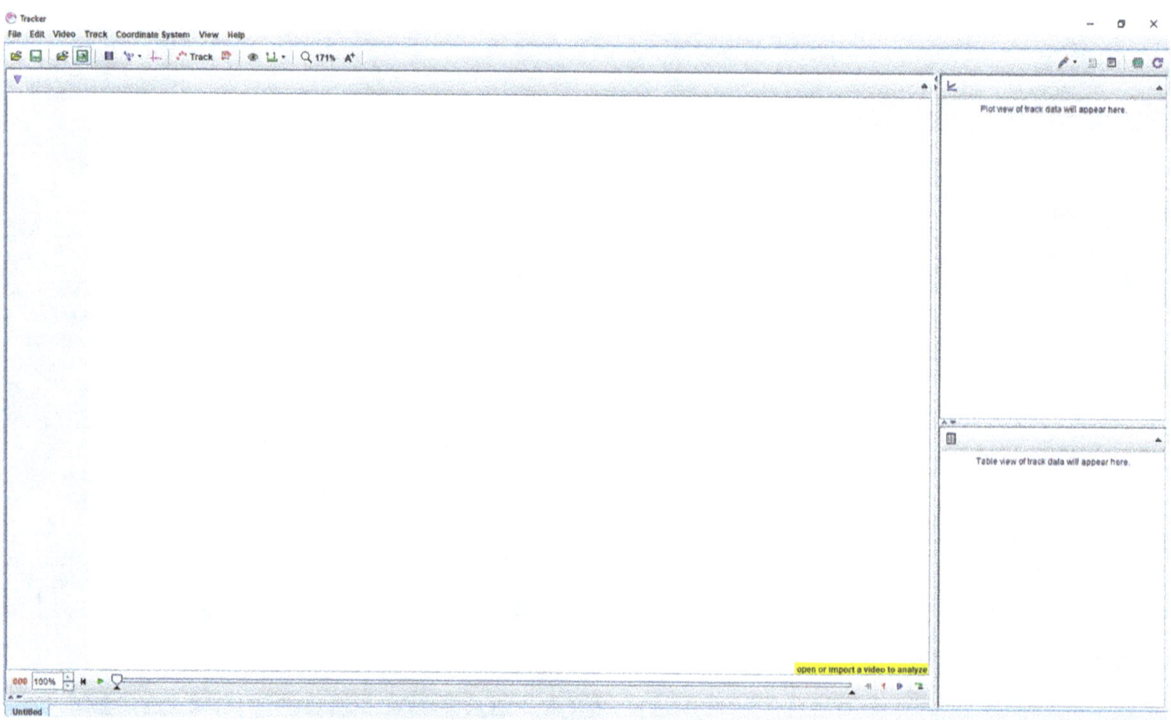

SETTING THE SCALE

The button to the left of the **Show or Hide the Coordinate Axes** has other controls. Using the drop down menu from this button, select **New**, and then select **Calibration Stick**. This will place a blue bar on the screen. Move each end of the bar to the reference length that you included in your video frame. You can click on the measurement on the calibration stick to update your reference length.

SETTING THE TIME

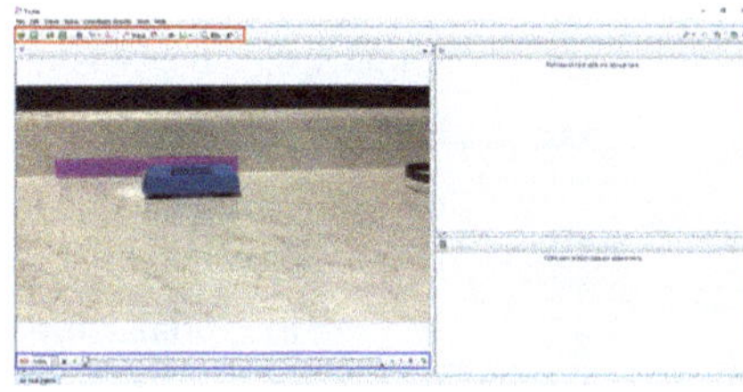

Notice in the image at right the video controls below the video outlined in blue. At the far left is the frame counter/time (depending on your setting). The bar and triangle will reset the video to its beginning. The green triangle starts and stops the video playback. The pentagon is a slider that you can use to navigate through the video. The blue bar and *left* triangle will reverse the video the number of frames indicated in red. The blue bar and *right* triangle will advance the video the number of frames indicated in red. The last control allows you to set the video to loop.

Advance the video to just prior to the event that you want to analyze. Once you have the video where you want it, click the frame counter and it will give you the option to set the time. You should set your time to zero.

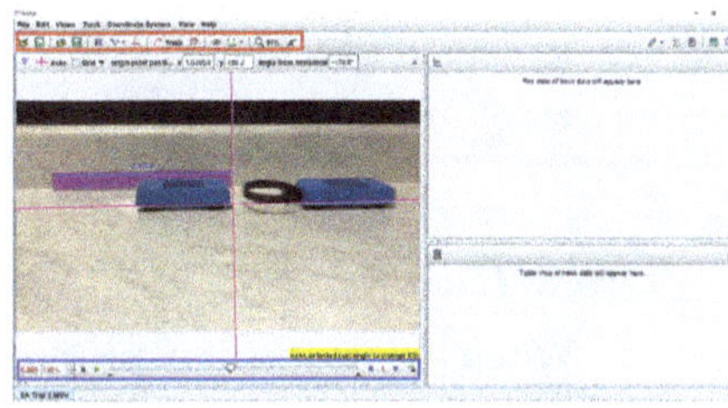

The image at right shows an example of an imported video in Tracker. You will see that we have set the coordinate axes and set the scale. In this example we had to slant the axes slightly, and the calibration stick was set to 0.300 m because it was the length that we used in our video. We have also set the start time for our event to just prior to the collision of the two carts.

Tracking the Motion

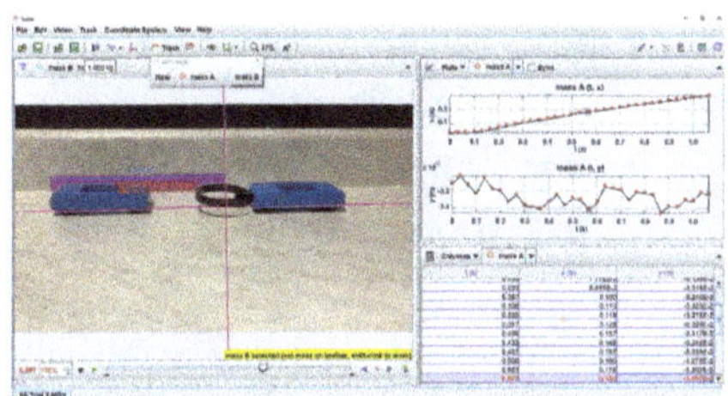

When you are ready to start collecting your data, on the upper menu click on **Track**, select **New**, and then select **Point Mass**. Decide which point on the object you will track. Select a point that will be clearly visible through the entire event. Hold shift and left click on the point that you will be tracking. As you click, the video should advance one frame. You can continue clicking your selected point until the event is over. As you click on each position, the data is entered into a position-time graph (upper right) and a data table (lower right). See the data track in the image at right. If you have additional objects to track, return the video to the start time and create a new point mass. Then plot the new positions. You will notice that the graph is not set to mass B, as it is in the data table. You can go back to mass A by using the drop down menus with the graphs and table.

Data Sets

While positions and times are great, we often want to analyze other dimensions. Notice the Columns drop-down menu above the data table. By selecting this, you can select a variety of data to include in your data table. For instance, you may want to include time, x-position, x-velocity, and x-acceleration. See right for an example using these parameters. You can analyze the graphs by right clicking on the graph, or you can copy all the data from your data table into a spreadsheet program.

Columns ▾ | mass A ▾

t (s)	x (m)	v_x (m/s)	a_x (m/s²)
0.000	4.251E-2		
0.033	4.449E-2	2.940E-2	
0.067	4.447E-2	2.881E-2	2.525
0.100	4.641E-2	0.191	3.409
0.133	5.723E-2	0.294	2.265
0.167	6.603E-2	0.308	0.365
0.200	7.779E-2	0.337	0.637
0.233	8.855E-2	0.338	-0.378
0.267	0.100	0.323	-0.504
0.300	0.110	0.293	-0.506
0.333	0.119	0.294	-0.493

APPENDIX E
Mathematical Reference

This appendix is provided as a ready reference for most of the mathematical principles required for this textbook. It is not intended to be exhaustive.

Mathematical Symbols

Symbol	Definition
$\equiv$	is defined as
$\ll, \gg$	is much less than, is much greater than
$\Leftrightarrow$	if and only if
$\Rightarrow$	implies; e.g., $a \Rightarrow b$ is read, "if *a*, then *b*"
$\propto$	is proportional to
$\pm$	plus or minus—value following may be positive or negative
Δx	change in *x*; final value minus initial value, or $\Delta x = x_f - x_i$
Σx	summation of *x*; the sum of all values represented by *x*
$x \to 0$	*x* approaches zero; the value of *x* becomes as small as we like
∞	infinity; the value is as large as we like

Determining Significant Figures

When we are reading measured data collected by someone else, we must be able to identify the significant figures (SF). Remember that significant figures do not apply to counts or definitions—only to measured data. Scientists follow an established set of rules to determine which digits in a measurement are significant.

Rule 1. All nonzero digits are significant.

Rule 2. All zeros between significant figures are significant.

Rule 3. All ending zeros to the right of the decimal point are significant.

Examples:

a. 1.694 mL $\Rightarrow$ 4 SF (all four digits are nonzero—Rule 1)

b. 51.021 m $\Rightarrow$ 5 SF (four nonzero digits and a zero between SF—Rule 2)

c. 22.7500 g $\Rightarrow$ 6 SF (four nonzero digits and two ending zeros to the right of the decimal point—Rule 3)

d. 0.097 N $\Rightarrow$ 2 SF (Rule 1. Since the zero in the tenths place is not an ending zero, Rule 3 does not apply.)

The fourth example can be confusing, but it is much easier to understand if we change the number to scientific notation.

$$0.097\ \text{N} = 9.7 \times 10^{-2}\ \text{N} \Rightarrow \text{only two SF}$$

One other case in which determining SF can be difficult is when ending zeros occur to the left of the decimal point.

Example:

How many SF does 7600 m have? We know that two of the digits must be significant, but we cannot determine which, if either, of the zeros are significant. Does this number have two, three, or four SF? Again, the person reporting the value can use scientific notation to avoid confusion.

$$7600\ \text{m} \Rightarrow 7.6 \times 10^3\ \text{m (two SF)}$$
$$7600\ \text{m} \Rightarrow 7.60 \times 10^3\ \text{m (three SF)}$$
$$7600\ \text{m} \Rightarrow 7.600 \times 10^3\ \text{m (four SF)}$$

ADDING AND SUBTRACTING DATA

Math Rule 1: Added or subtracted data must have the same units. Even if the kinds of data are the same, if their units are not the same, then we are adding apples and oranges. For example, we can't add the lengths 3.1 m and 45 cm together without first converting one of the measurements to the other's units.

Math Rule 2: The result of adding or subtracting data can't be more precise than the least precisely measured data used in the calculation. After adding or subtracting, round the result to the decimal place of the estimated digit in the least precise measurement.

Example 1 Adding Data: Length

Add 3.1 m and 45 cm.

Unit Conversion:

$$45\ \cancel{\text{cm}}\left(\frac{1\ \text{m}}{100\ \cancel{\text{cm}}}\right) = 0.45\ \text{m}$$

Add data in column:

$$\begin{array}{r} 3.\underline{1}\ \text{m} \\ +\ 0.4\underline{5}\ \text{m} \\ \hline 3.\underline{55}\ \text{m} = 3.6\ \text{m} \end{array}$$

The estimated digits in each piece of data and in the sum are underlined. The sum contains two underlined digits that result from the summing operation. These must be rounded to the place having the least precise estimated value, in this case the 0.1 m place.

Multiplying and Dividing Data

Math Rule 3: The result of multiplying or dividing data can't have more SF than the data with the fewest SF used in the calculation. After multiplying or dividing, round the results to the same number of significant figures as the data with the fewest SF.

Example 2 Dividing Data: Density

Calculate the density of a sample of quartz with a mass of 27.55 g and a volume of 10.4 cm³.

Given data:

$m = 27.55$ g (four SF), $V = 10.4$ cm^3 (three SF)

Calculation:

$$d = \frac{m}{V}$$

$$= \frac{27.55\ \text{g}}{10.4\ \text{cm}^3}$$

$= 2.65$ g/cm^3 (three SF allowed)

Since 10.4 cm^3 has only three SF, only three SF are allowed in the quotient.

Math Rule 4: The result of multiplying or dividing a measured quantity and a pure number has the same number of decimal places, or the same precision, as the measured quantity used in the calculation.

Examples:

a. (7)(2.35 cm) = 16.45 cm (not 16.5 cm)

b. 2.63 cm ÷ 5 = 0.53 cm (not 0.526 cm)

Notice that the number of SF in the measured data does not determine the number of SF in the result when multiplying or dividing by pure numbers. Math Rule 4 ensures that we preserve the precision of the original measurement, even if we change the number by multiplying or dividing by a pure number.

Multistep Calculations

Math Rule 5: When doing multistep calculations, we do not round intermediate-step answers. When we get to each intermediate step, we keep one extra digit after the least significant digit and truncate (delete without rounding) the remaining digits (these digits are uncertain).

Example 3 Multistep Calculations: Density

Calculate the average density of two samples of iron with masses of 29.42 g and 44.9715 g and volumes of 3.57 cm^3 and 5.6859 cm^3 respectively.

Given data:

$m_1 = 29.42$ g (four SF),
$m_2 = 44.9715$ g (six SF),
$V_1 = 3.57$ cm^3 (three SF),
$V_2 = 5.6859$ cm^3 (five SF)

Calculation:

$$d = \frac{m}{V} = \frac{m_1 + m_2}{V_1 + V_2}$$

$$= \frac{29.42\ \text{g} + 44.9715\ \text{g}}{3.57\ \text{cm}^3 + 5.6859\ \text{cm}^3}$$

$$= \frac{74.3\underline{9}15\ \text{g}}{9.2\underline{5}59\ \text{cm}^3}$$

Keep one extra digit and truncate.

$$= \frac{74.3\underline{9}1\ \text{g}}{9.2\underline{5}5\ \text{cm}^3} \left(\frac{\text{four SF}}{\text{three SF}}\right)$$

$= 8.04$ g/cm^3 (three SF allowed)

Rounding Rules

We round numbers all the time. Sometimes it's convenient to round, for example, when referring to the number of residents in your town to the nearest 100 people or to the distance around the earth to the nearest 1000 kilometers. Rounding is useful for several reasons. It helps us grasp large numbers when the specific number is not needed but the approximate size is. Rounding also allows us to report measured scientific data correctly.

APPENDIX E
Mathematical Reference

For the purposes of measuring scientific data and performing calculations, scientists have agreed on certain rounding rules. These ensure that calculation solutions are properly rounded to show the needed precision of numerical quantities. We suggest that you use the following rules for problems in this textbook to closely imitate those used by scientists.

1. Identify the place value that you are going to round to. This is the rounded place value.
2. If the digit to the right of the rounded place value is 0–4, the digit in the rounded place value remains unchanged ("rounded down"). If the digit to the right is 5–9, the digit in the rounded place value is increased by 1 ("rounded up").
   a. If the rounded place value is to the *right* of the decimal point, all digits to the *right* of the rounded place value will be dropped after rounding.

      Examples: $105.\underline{6}39$ g ⇒ 105.6 g

      $82.\underline{4}7$ mm ⇒ 82.5 mm

      $32.\underline{9}5$ cm ⇒ 33.0 cm

   b. If the rounded place value is to the *left* of the decimal point (or assumed decimal point), all digits between the rounded place value and the decimal point become zeros and the decimal point (if one is shown) is omitted.

      Examples: $1\underline{6}94$ m ⇒ 1700 m

      $1\underline{9}.6$ °C ⇒ 20 °C

3. Notice in two of the examples above that rounding to a particular place value may affect other digits to the left of the rounding place value. When a 9 is rounded up, the 9 becomes 0 and 1 is added to the digit to the left—carried to the next decimal place.

Scientific Notation

1. Scientists and engineers often deal with extremely large or small numbers. To make working with such numbers easier, a shorthand method of writing them is used, called *scientific notation.*
2. In scientific notation the number is written in the form $M \times 10^n$. M can be positive or negative but must be greater than or equal to 1 and less than 10 ($1 \le |M| < 10$). The exponent, n, is an integer and can be positive or negative as well.
3. To convert a number to scientific notation, move the decimal point left or right as necessary until only one nonzero digit appears to the left of the decimal point. The number of places that the decimal point has moved will become the exponent (n) in your final expression. If you moved the decimal to the left, the exponent will be positive. If you moved the decimal point to the right, the exponent will be negative. Retain all the significant figures present in the original number in M. All nonsignificant zeros should be dropped. See the table below for several examples.

Decimal Form	Scientific Notation
34 500 000	3.45×10^7
176.00	1.7600×10^2
0.000 057	5.7×10^{-5}

Notice which instances in the table the zeros remain in the scientific notation form. In the first and third entries the zeros are not retained because they are not significant. In the second entry the zeros are retained because they are significant.

ADDITION AND SUBTRACTION IN SCIENTIFIC NOTATION

Numbers written in scientific notation must have the same powers of 10 before they can be added or subtracted. If the exponents are not the same, convert one of the numbers before you perform the operation. Converting the exponent term of the smaller number to the larger exponent is usually a good practice. Report the result in proper scientific notation.

Example 4 Addition and Subtraction in Scientific Notation

Add 3.5×10^4 kg and 7.3×10^3 kg.

1. Convert one of the numbers so that they have the same exponent.

   $7.3 \times 10^3 \text{ kg} = 0.73 \times 10^4 \text{ kg}$

2. Add, following the rules for significant figures, if applicable.

$$\begin{array}{r} 3.5 \times 10^4 \text{ kg} \\ +\ 0.73 \times 10^4 \text{ kg} \\ \hline 4.\underline{2}3 \times 10^4 \text{ kg} \end{array}$$

3. If necessary, convert answer to proper scientific notation.

   4.2×10^4 kg

MULTIPLICATION AND DIVISION IN SCIENTIFIC NOTATION

Multiplication and division can be done on numbers in scientific notation without changing their form. Perform the operation on the decimal (M) factors first. Then multiply or divide the exponential terms following the rules for exponents. The rules for exponents are summarized at the bottom of the next column. Convert the resulting number to standard scientific notation.

Example 5 Multiplication and Division in Scientific Notation

a. $(2.5 \times 10^7\ \text{m})(1.4 \times 10^{-3}\ \text{m})$

$$= (2.5)(1.4)(10^7 \times 10^{-3})\ \text{m}^2$$

$$= 3.5 \times 10^{7+(-3)}\ \text{m}^2$$

$$= 3.5 \times 10^4\ \text{m}^2$$

b. $\dfrac{2.7 \times 10^5\ \text{m}}{3.0 \times 10^{-2}\ \text{s}} = \left(\dfrac{2.7}{3.0}\right) \times \left(\dfrac{10^5}{10^{-2}}\right)\left(\dfrac{\text{m}}{\text{s}}\right)$

$$= 0.90 \times 10^{5-(-2)}\ \text{m/s}$$

$$= 0.90 \times 10^7\ \text{m/s}$$

$$= 9.0 \times 10^6\ \text{m/s}$$

Logarithms and Exponential Operations

1. Any positive real number n can be expressed as a base number, a, raised to some exponent, b, and can be written as $n = a^b$.
2. The number b is called the *logarithm* of n to the base a, so a logarithm is nothing more than an exponent. This expression is written as $\log_a n = b$.
3. Any positive number may be the base of a logarithm, but you are likely to come across only certain numbers used as bases in physics. Logarithms of base 10 are called *common logarithms,* while those of the base e (2.718 28 . . .) are called *natural logarithms.* The common log of n is expressed as $\log n = b$, where the base is understood to be 10. The natural log of n is expressed as $\ln n = b$, where the base is understood to be e.
4. The logarithm of numbers greater than 1 are positive decimal values. The logarithms of numbers between 0 and 1 are negative decimals. The logarithm of 1 is zero (0). We cannot obtain a logarithm for zero nor for a negative number.
5. The significant figures to the right of the decimal point matches the number of significant figures in the value for which we take the logarithm.
   - **a.** $\log 25 = 1.40$ (two SF in 25, therefore, two SF to the right of the decimal point)
   - **b.** $\log 0.450 = -0.347$ (three SF in 0.450, therefore, three SF to the right of the decimal point)
   - **c.** $\ln 315 = 5.753$ (three SF in 315, therefore, three SF to the right of the decimal point)
   - **d.** $\ln 0.9725 = -0.027\,89$ (four SF in 0.9725, therefore, four SF to the right of the decimal point)

RULES OF LOGARITHMS

a. $\log_a a = 1$

b. $\log 1 = 0$

c. $\log (ab) = \log a + \log b$

d. $\log \frac{a}{b} = \log a - \log b$

e. $\log a^b = b \log a$

f. $\log \sqrt[b]{a} = (\log a) \div b$

RULES OF EXPONENTS

a. $x^1 = x$

b. $x^0 = 1$

c. $\frac{1}{x^a} = x^{-a}$

d. $x^a x^b = x^{a+b}$ (bases must be the same)

e. $\frac{x^a}{x^b} = x^{a-b}$ (bases must be the same)

f. $\sqrt[a]{x} = x^{\frac{1}{a}}$

APPENDIX E
Mathematical Reference

Geometric Figures

Two-Dimensional Figures

Figure	Perimeter	Area
Circle	$2\pi r$	πr^2
Triangle	—	$\frac{1}{2}bh$
Rectangle	$2(l + w)$	lw

Three-Dimensional Figures

Figure	Surface Area	Lateral Area	Volume
Sphere	$4\pi r^2$	N/A	$\frac{4}{3}\pi r^3$
Rectangular Prism	$2(lw + wh + lh)$	$2h(l + w)$	lwh
Right Circular Cylinder	$2\pi rh + 2\pi r^2$	$2\pi rh$	$\pi r^2 h$
Right Circular Cone	$\pi r^2 + \pi r\sqrt{r^2 + h^2}$	$\pi r\sqrt{r^2 + h^2}$	$\frac{1}{3}\pi r^2 h$

Functions

One aspect of God's creation that man discovered is that many phenomena can be modeled with simple equations that relate a physical quantity to one or more others. For example, if we know the speed of a car and how long it has been traveling at that speed, we can tell how far the car traveled during that time interval. Such relationships are called *functions* because a set of input numbers (the independent variable set) produces a unique set of output numbers (the dependent variable set).

We will study a variety of functions in physics. In this course they will be limited to a few simple types.

LINEAR FUNCTIONS

Linear functions are the simplest type and produce a straight-line graph. Linear functions may be written in the form

$$y = mx + b,$$

where m is the slope of the line and b is the y-intercept. If the relationship between the independent and dependent variables is direct, then the y-intercept is zero ($b = 0$) and the line passes through the origin of the graph (0,0).

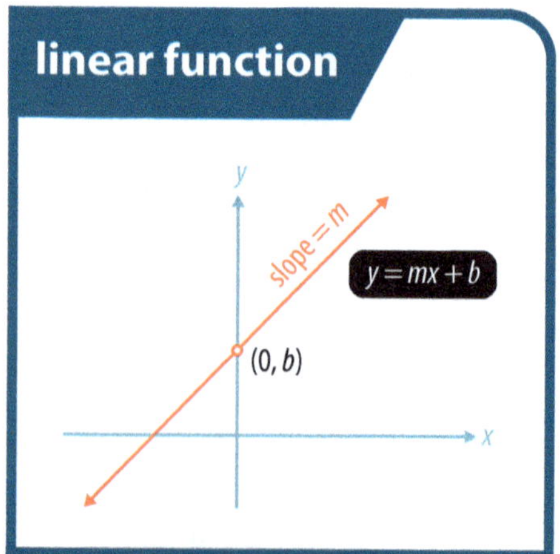

linear function

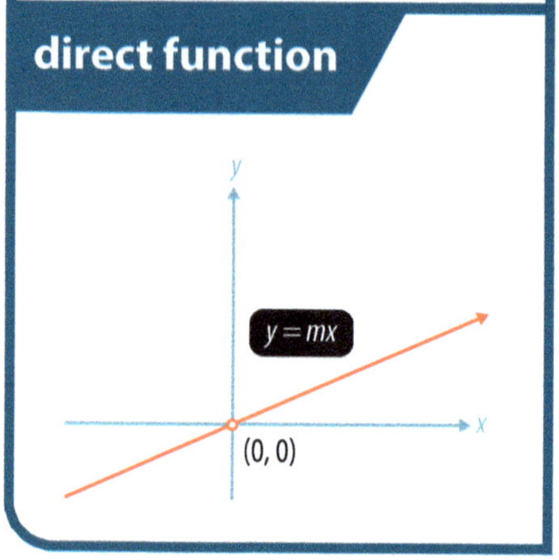

direct function

QUADRATIC FUNCTIONS

Quadratic functions are relationships where the dependent variable is a function of the square of the independent variable. Graphs of quadratic functions are parabolas. Quadratic functions are written in the form

$$y = ax^2 + bx + c,$$

where a, b, and c are constants.

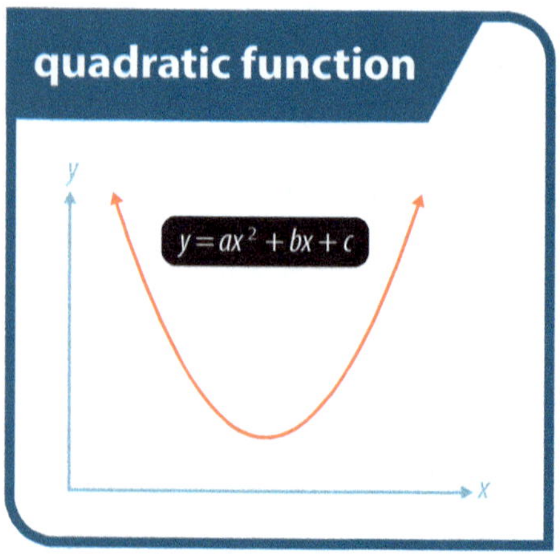

quadratic function

INVERSE FUNCTIONS

Inverse functions are relationships where, as the independent variable increases, the dependent variable decreases in a reciprocal fashion. Such relationships are called *inverse proportions*. Graphs of inverse functions are hyperbolas. These functions may be written in the form of either

$$y = \frac{k}{x} \text{ or } yx = k,$$

where k is a proportionality constant. Boyle's law provides an example of an inverse function. Note that neither variable ever reaches zero nor crosses its axis in the inverse function of a physical phenomenon.

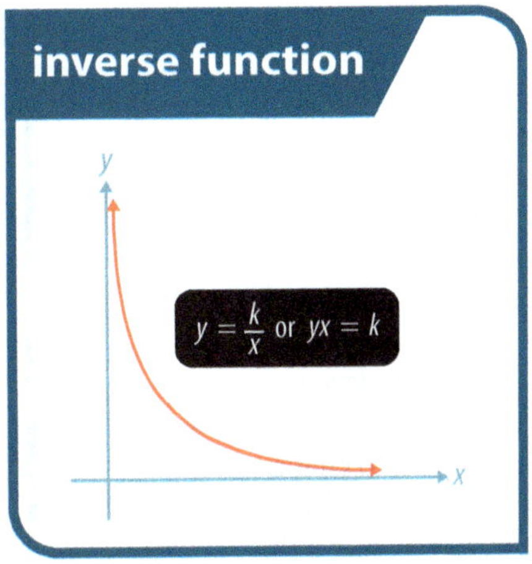

EXPONENTIAL FUNCTIONS

Exponential functions are very common in natural processes. In exponential functions the dependent variable is the result of some base number raised to the power equal to the independent variable. These functions can be written in the form of either

$$y = n^x \text{ or } y = n^{-x},$$

where n is a positive real number. The magnitude of n determines how steep or how shallow the curves are. The graphs of exponential functions form sweeping rising (exponential growth) or falling (exponential decay) curves (see right).

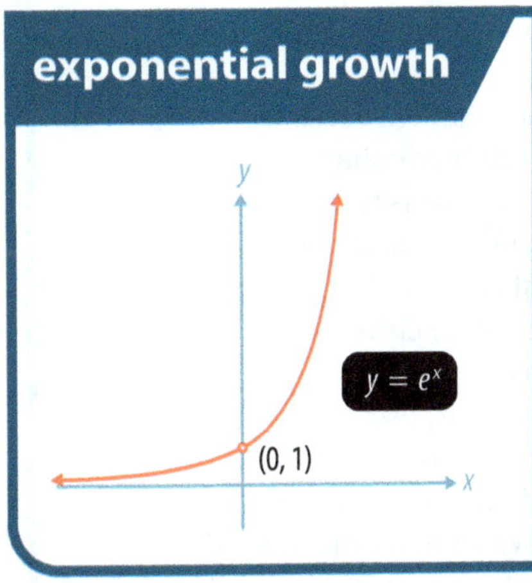

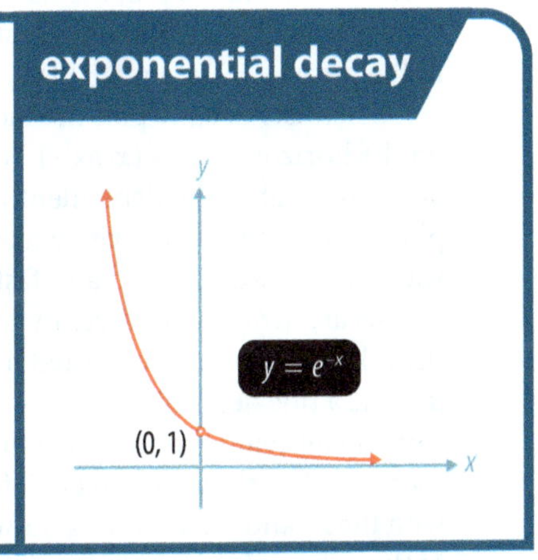

PERIODIC FUNCTIONS

Periodic functions are functions that repeat or that show an oscillation of the dependent variable over the domain of the independent variable. Such functions include the trigonometric functions. The equation form of these functions varies depending on the operator and argument but generally follows the form

$$y = A\ op(\theta + k) + c,$$

where A, k, and c are all constants, and the operator op may be any periodic function operator. The quantity within the parentheses may be an angle or an angular frequency that is dependent on time. See several graphs of typical periodic functions at right.

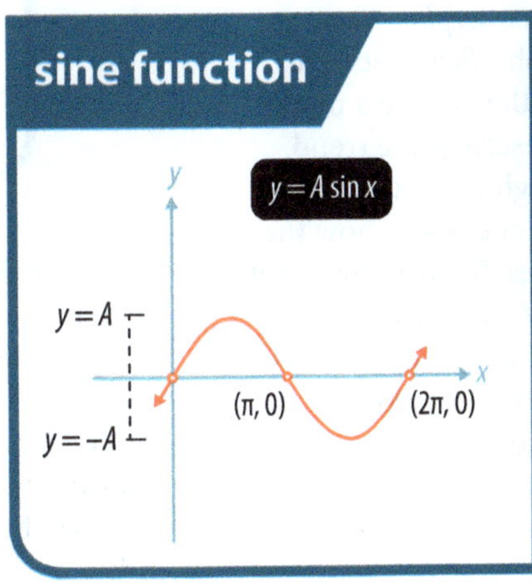

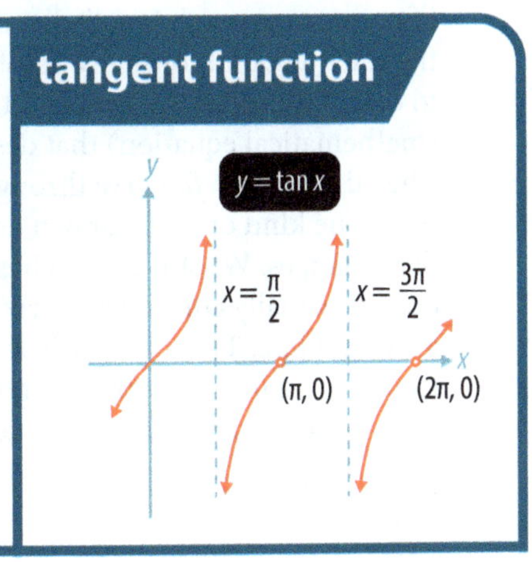

APPENDIX E
Mathematical Reference

Graphing

Scientists often like to compare two or more groups of numbers visually to see whether they are related in some way. When scientists plot data to compare different quantities, they make graphs. The simplest graphs compare two changing quantities, called *variables*. Usually, one of the variables is determined by the scientist or else changes in a regular way (e.g., time). This is called an *independent variable*. Its value doesn't depend on anything in the data. The other variable is expected to change in some way related to the independent variable. Its value depends on the first variable, so it is called the *dependent variable*. The pair values of the independent and dependent variables are called the *coordinates* of the data. To plot the data, an ordered pair of coordinates is used. You have graphed ordered pairs in the form (x, y) in a math class.

Scientists usually plot the independent variable on the horizontal axis (x-axis), with increasing values to the right. The dependent variable is typically plotted on the vertical axis (y-axis), increasing upward. These are not hard and fast rules, and many graphs are arranged differently to more clearly see the relationships in the plotted data. Useful graphs include a title describing the graph's purpose and labels identifying the quantities and units used on each axis. Graphs are often depicted on a scaled grid with the x- and y-axis scales shown so that estimates of the values of the plotted variables can be made.

SCATTERPLOTS

Graphs come in different forms. A simple plot of points on a graph is called a *scatterplot*. This is the starting point for many graphs. Scientists like to detect trends in the data and then create a model (mathematical equation) that describes the trend. They draw a *best-fit curve* through the pattern of dots. The kind of curve drawn depends on how the data changes. We still call it a best-fit curve even if it doesn't actually curve. The curve of a *linear graph* is a straight line. These are fairly rare in nature. Most trends in nature range from slightly curved to really wavy! These graphs, logically enough, are called *nonlinear graphs*.

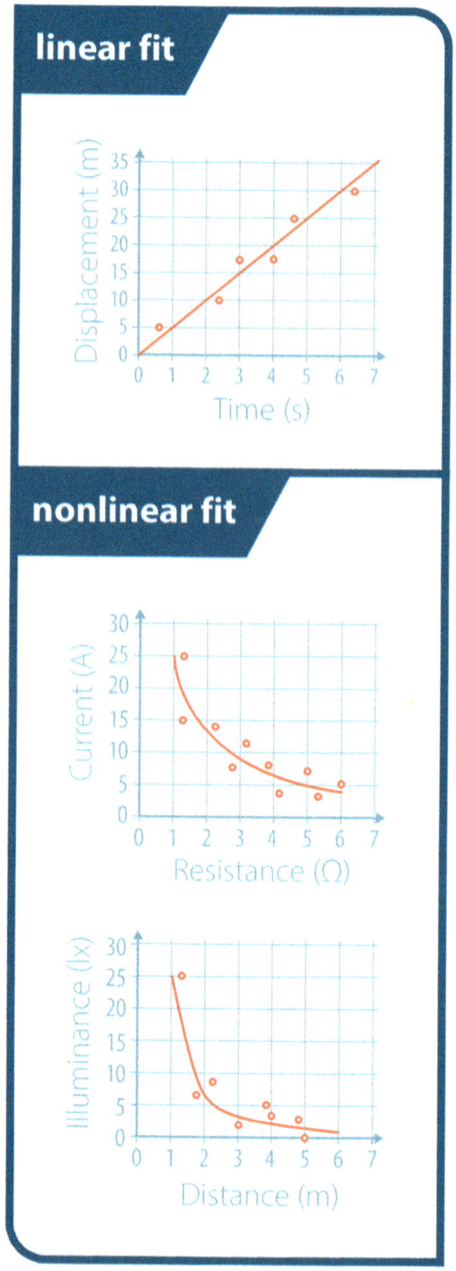

Since the purpose of the scatterplot is to determine a mathematical model for the data, scientists need a tool to compare their model to the data. A common method of assessing the fit of a particular model to the data is the coefficient of determination, r^2or R^2. This number indicates what percentage of the change in the y-values is caused by changes in the x-value. A high r^2 value indicates that the model is a "good fit" of the data; that is, the model predicts y-values with reasonable accuracy. Low values of r^2 indicate a "poor fit" of the model to the data. As we work to fit models to a set of data, the model that produces the highest r^2 value is often the best model of the data. We must also consider the implications of a particular model. For example, a quadratic which is an even function will reverse direction at some point; we know that the tails of a quadratic point in the same direction. Therefore, a quadratic model is predicting a change in direction at some point. We must consider whether that makes sense for the physical situation being modeled. For instance, this does make sense for projectile motion because we know that most things that we throw upward will come back down.

SLOPE

One of the most important things to be learned from a graph is the rate at which the dependent variable changes in comparison to the independent variable. This rate is shown by the slope of the graph. A graph with a steep slope shows that things are changing quickly. One with a horizontal slope shows that the dependent variable is not changing at all. A graph that rises from left to right has a positive slope; a dropping curve has a negative slope.

We can calculate the slope of a line, but we can also determine the slope of any curve by calculating the slope of a tangent to the curve at that point. To calculate the slope of a curve, use the slope formula,

$$m = \frac{\Delta y}{\Delta x} = \frac{y_2 - y_1}{x_2 - x_1},$$

where m is the slope, Δy is the change in the y variable, and Δx is the change in the x variable.

ESTIMATING DATA

We can sometimes use scatterplots and trend lines to obtain information that is not in the measured data set. If we follow the trend line between two data points, these values are estimated—not measured. Obtaining unmeasured data between collected data points like this is called *interpolation*. Scientists also often try to predict the values of the dependent variable beyond the range of the measured independent variable data points. To estimate data outside the range of collected data points is known as *extrapolation*. Extrapolating data assumes that the trend will continue in the same fashion as observed within the measured data.

APPENDIX F
STEM Design Process

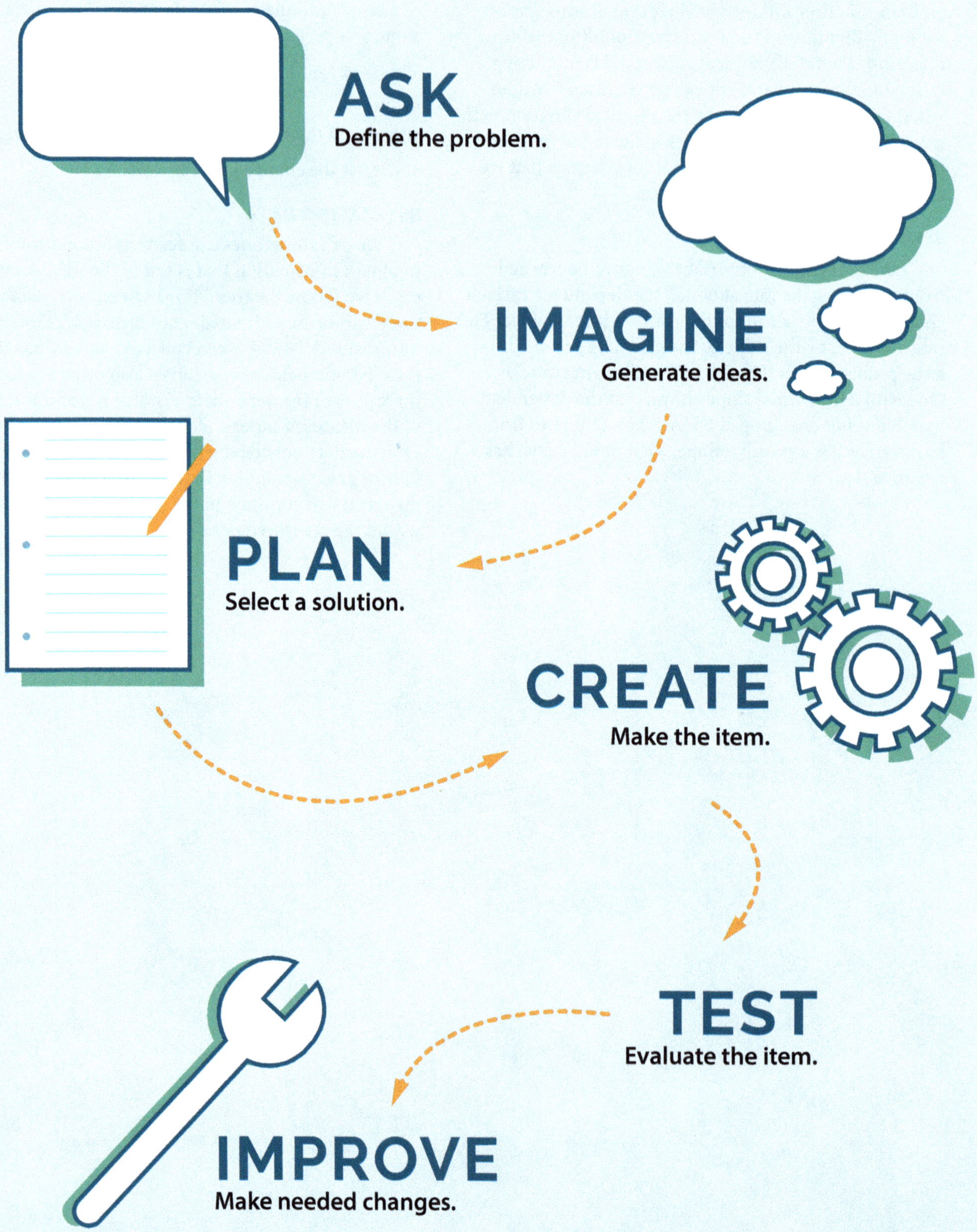

APPENDIX G
Writing Formal Lab Reports

As you have learned in your study of science, communication is an important part of the scientific process. Part of what scientists do is share what they have learned with other scientists. One example of this is the writing of formal lab reports. In a lab report a scientist presents a problem that was investigated, describes how the investigation was conducted, and summarizes what was learned during the investigation. Many college science courses require students to present all their laboratory work in formal lab reports.

To better prepare you for college, your teacher may choose to have you submit your lab assignments in formal lab reports rather than use the fill-in-the-blank format found in your lab manual. You may be more familiar with the latter form, but there is really little to be afraid of about writing formal reports. They may be new to you, and the word "formal" may sound alarming, but with practice such reporting can become second nature.

There really isn't a standardized way of writing and compiling a formal lab report. Your teacher may require you to do all your reporting in a bound journal, or you may be asked to submit individual typed reports. Regardless of which method your teacher prefers, there are some properties of lab reports that teachers generally agree upon. Let's look at the style and mechanics of a formal lab report and the type of content that is normally included.

Style and Mechanics

The first thing to keep in mind about a formal lab report is that it is a type of *formal writing*. That means that a formal lab report is like the final draft of an essay or research paper. It should have a neat, orderly appearance. To that end, formal lab reports normally have the following elements of style and mechanics.

- Unbound lab reports should be typed in black ink. Reports in a journal should be *neatly* written.
- Lab reports should be written in complete sentences and be checked for proper spelling, punctuation, and grammar.
- Formal lab reports should be free of erasures, whiteout, or scratched-out words.
- Since objectivity is a key aspect of the scientific process, lab reports usually avoid first-person narrative and instead use an impersonal third-person form.

Example:

DON'T SAY: "I learned that 10 mL of *A* will neutralize 5 mL of *B*." (less formal)

DO SAY: "The results indicated that 10 mL of *A* will neutralize 5 mL of *B*." (more formal)

Format

Scientists (and science teachers) may disagree on the exact number of components in a formal lab report, but there is a general consensus on what is presented and the order in which it is presented. Keep in mind as you write that a formal lab report answers three basic questions.

- What problem was investigated?
- What procedure was used during the investigation?
- What was learned from the investigation?

To answer those questions, a formal lab report generally includes the following elements:

- title
- synopsis
- introduction
- list of any materials used and a description of the procedure
- data yielded by the procedure
- analysis of the data and a discussion of the findings
- sources

Let's take a closer look at each of these.

Title

A formal lab report title should succinctly describe what the lab activity was about. It should be matter-of-fact rather than creative. Consider the following possible titles for a titration experiment.

Example:

INFORMAL: Drip, Drip, Drip!

FORMAL: Comparing the Strengths of Acids Using Titration

It's OK for a formal lab report title to sound dry—save the creative writing for your language arts class! Imagine yourself as another scientist who is reading your report as part of her research; she wants to know right away what your report is about without having to read deep into the main body of the text.

Synopsis

A synopsis is a very brief summary of the entire lab report, that is, two or three sentences. If the title of your report catches the reader's attention, the synopsis should then give the reader just enough detail for him to decide whether he wants to read further.

Example (based on a titration lab):

The suitability of titration using an aqueous solution of sodium hydroxide (NaOH) for distinguishing between acids of different strengths was tested. The results confirmed the ability of sodium hydroxide solution to distinguish between different acids on the basis of a direct relationship between the amount of sodium hydroxide required to neutralize an acid and the number of hydronium ions produced by the acid in an aqueous solution.

Introduction

Your introduction should expand on your synopsis by giving more details about the experiment. An introduction should include the following elements:

- a description of the purpose of the experiment
- the hypothesis that will be tested
- a summary of the methods that will be used to test the hypothesis (Be brief. A more detailed description should be reserved for the Materials and Methods section that follows.)
- a brief description of the results of the procedure
- a summary of the conclusions that you have drawn on the basis of the results

Materials and Methods

The goal of this section of your lab report is to describe your investigation in enough detail to allow another person to assess its validity or even replicate it. That is why there are two parts to this section: materials *and* methods. You should first list your *materials*, including names and quantities where appropriate (e.g., for chemicals). Your list should be just that—a list. Think about how you normally see such lists in a student lab manual.

Next, describe your *methods*. You do not need to give instructions for how to do common laboratory tasks, such as using an electronic balance. Nor do you need to explain how to do common laboratory procedures if your audience is expected to be familiar with them. For example, in the titration example used above it would not be necessary to explain how a titration is done; your audience probably already knows how to do this. What your audience *would* need to know in order to replicate your experiment is the chemicals that you used along with how much and in what concentration. As you write this section, ask yourself whether someone reading your descriptions would be able to replicate your experiment exactly. If the answer is no, then you need to be more specific. It is often helpful to include a diagram of any experimental setup used.

Data

Your data section should be reserved for raw data only. This is where you report the results of your tests. Your data may be *qualitative* (i.e., based on observations) or *quantitative* (i.e., based on measurements). Often, a good way to report quantitative data is to use a data table. The format of data tables will vary depending on the data that you collected. An excellent technique, especially if you are using spreadsheets, is to record the units of measure in the column header and record only numerical data in the cells. This also allows you to use the computational features of the software. All quantitative data should be reported with the appropriate number of significant figures to indicate the precision of the measurements.

Analysis and Discussion

This is where the fun part begins! Now you get to interpret the data that you presented in the previous section. First, describe how the data was analyzed and show any formulas or equations that were used to do so. Sample calculations should be included. Calculated values should be reported with the proper number of significant figures. Descriptions of the results should include appropriate statistical parameters, such as the mean, median, and range of values. Where appropriate, use graphs to show trends in the data. All graphs should be properly scaled and include titles and axes labels with units.

Once you have presented the analysis of your data, you may then address the all-important question: What does it mean? Think back to your stated hypothesis—did your data and analysis support your hypothesis? Were your results in line with what you expected? How did your results compare with those of others in your class? Be sure to support any claims you make with the relevant data and analyses. Your reader—your teacher—expects to see logical arguments that are consistent with your experimental results.

Your discussion needs to be frank and honest. Don't try to make your data say something that it isn't actually saying. If your data does not support your hypothesis, then say so! Don't say that it does just because you think that might be the answer that your teacher is expecting. An important aspect of nurturing science skills is learning how to discuss inconclusive or contrary data. Examine what may have gone wrong during the experiment, make suggestions for improving future investigations, or propose a modified hypothesis for future testing.

It is also appropriate during your discussion to suggest possible practical applications of your work or to put forward additional related questions that might be investigated in the future. In doing so, you might be laying the groundwork for a future scientist. By this you can see how the process of science has links both to those who have investigated a question in the past and to those who may investigate it in the future.

Sources Cited

Much like any formal research paper, a lab report should cite sources within the body of the report and include a bibliography at the end. There are different formats for doing this. Your teacher will advise you on what format to use.

APPENDIX H
Physical Constants

Fundamental Constants

Quantity	Symbol	Value
atomic mass unit	u	$1.660\,539\,066\,60 \times 10^{-27}$ kg
Avogadro constant	N_A	$6.022\,140\,76 \times 10^{23}$/mol
Boltzmann constant	k_B	$1.380\,649 \times 10^{-23}$ J/K
Coulomb constant	k	$8.987\,551\,79 \times 10^{9}$ N·m²/C²
electron mass	m_e	$9.109\,383\,70 \times 10^{-31}$ kg
fundamental charge	e	$1.602\,176\,634 \times 10^{-19}$ C
gas constant	R	8.314 462 6 J/mol·K
gravitational constant	G	$6.674\,301\,5 \times 10^{-11}$ N·m²/kg²
neutron mass	m_n	$1.674\,927\,5 \times 10^{-27}$ kg
permeability of free space	μ_0	$1.256\,637\,062 \times 10^{-6}$ T·m/A
permittivity of free space	ε_0	$8.854\,187\,812 \times 10^{-12}$ C²/(N·m²)
Planck constant	h	$6.626\,070\,15 \times 10^{-34}$ J·s
reduced Planck constant	$\hbar$	$1.054\,571\,817 \times 10^{-34}$ J·s
proton mass	m_p	$1.672\,621\,92 \times 10^{-27}$ kg
speed of light in vacuum	c	$2.997\,924\,58 \times 10^{8}$ m/s
Stefan-Boltzmann constant	σ	$5.670\,374\,419 \times 10^{-8}$ W/(m²·K⁴)
Wien displacement law constant	b	$2.897\,771\,96 \times 10^{-3}$ m·K

Other Useful Constants

Quantity	Symbol	Value
absolute zero	0 K	−273.15 °C
astronomical unit	au	$1.495\,978\,7 \times 10^{8}$ km
atmospheric pressure, standard	—	101.325 kPa
density of dry air	—	1.20 kg/m³ (20 °C and 101.325 kPa)
density of water	—	1.000×10^{3} kg/m³ (4 °C)
gravitational acceleration, standard	g	9.807 m/s² (average, near Earth's surface)
latent heat of fusion of water	L_f	3.34×10^{5} J/kg
latent heat of vaporization of water	L_v	2.256×10^{6} J/kg
light year	ly	9.4607×10^{12} km
lunar gravitational acceleration	g_{moon}	1.62 m/s²
parsec	pc	3.0857×10^{13} km
rest energy of 1 u	—	931.5 MeV
speed of sound in air	—	343 m/s (20 °C and 101.325 kPa)
triple point of water	273.16 K	0.01 °C (at 611.657 Pa)

APPENDIX I
Electrical Resistor Color Codes

The following table is applicable for most common four- and five-color resistor codes. Five-color resistors are usually precision resistors with tighter tolerances (±1% or ±2%). Four-color resistors are usually lower tolerance (±5%, ±10%, or ±20%).

Resistor Color Codes

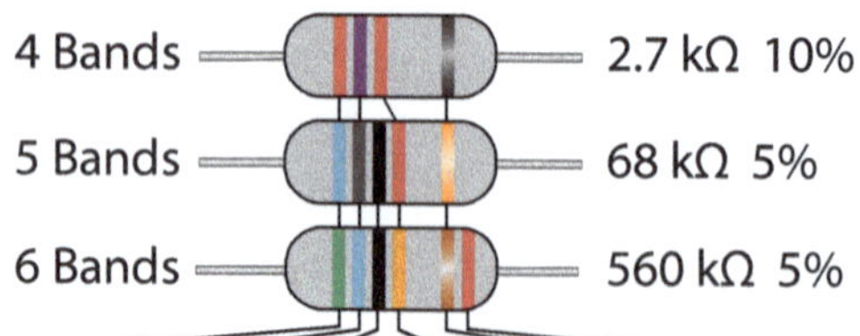

1st Digit	2nd Digit	3rd Digit	Multiplier	Tolerance	Temperature Coefficient
0	0	0	1		
1	1	1	10	1%	
2	2	2	100	2%	
3	3	3	1 k		
4	4	4	10 k		
5	5	5	100 k	0.5%	
6	6	6	1 M	0.25%	
7	7	7	10 M	0.1%	
8	8	8	0.01	10%	
9	9	9	0.1	5%	

The color bands printed on the end or middle of the resistor indicate the resistance. The first two bands (or three bands for five-color resistors) are the significant figures of the resistance. The total resistance is determined by multiplying the two-digit (or three-digit) number by the multiplier indicated by the next band's color. The last color band indicates the range of resistance that may be observed under actual use (i.e., the resistor's tolerance). For small resistors it may be difficult to identify the end containing the significant figures. If the first band appears to be a gold or silver color, then start from the opposite end, since these colors are always reserved for multiplier or tolerance values. See the sample resistor below.

FOUR-BAND RESISTOR

First Band	(first SF)	red	2
Second Band	(second SF)	violet	7
Third Band	(multiplier)	orange	1 K
Fourth Band	(tolerance)	gold	±5%

The resistance is (27 Ω × 1 K) ±5% = 27 kΩ ± 5%, or a range of 25.7 kΩ to 28.4 kΩ.

FIVE-BAND RESISTOR

First Band	(first SF)	brown	1
Second Band	(second SF)	green	5
Third Band	(third SF)	black	0
Fourth Band	(multiplier)	black	1
Fifth Band	(tolerance)	red	±2%

The resistance is (150 Ω × 1) ±2% = 150 Ω ± 2%, or a range of 147 Ω to 153 Ω.

APPENDIX J

Characteristics of Selected Subatomic Particles[1]

Family Name		Particle Name	Particle Symbol	Antiparticle Symbol	Rest Energy (MeV)	Particle Lifetime
Lepton		electron	e^-	e^+	0.511	$> 6.6 \times 10^{28}$ y
		muon	μ^-	μ^+	105.7	2.20×10^{-6} s
		tauon	τ^-	τ^+	1777	2.90×10^{-15} s
		electron neutrino[2]	ν_e	$\bar{\nu}_e$	$< 1.1 \times 10^{-6}$	
		muon neutrino[2]	ν_μ	$\bar{\nu}_\mu$	< 0.19	
		tauon neutrino[2]	ν_τ	$\bar{\nu}_\tau$	< 18.2	
Hadron	Meson	Pion (light, unflavored)	π^-	π^+	139.6	2.6×10^{-8} s
			π^0	π^0	135.0	8.4×10^{-17} s
		Kaon (strange)	K^+	K^-K^-	493.7	1.2×10^{-8} s
			K^0	$\bar{K}^0$	497.6	$\approx 10^{-9}$ s note 3
	Baryon	proton	p	$\bar{p}$	938.3	$> 10^{29}$ y note 3
		neutron	n	$\bar{n}$	939.6	879.4 s
		Lambda	Λ	$\bar{\Lambda}$	1116	2.6×10^{-10} s
		Sigma	Σ^+	Σ^-	1189	0.8×10^{-10} s
			Σ^0	$\bar{\Sigma}^0$	1193	7.4×10^{-20} s
			Σ^-	$\bar{\Sigma}^+$	1197	1.5×10^{-10} s
		Xi	Ξ^0	$\bar{\Xi}^0$	1315	2.9×10^{-10} s
			Ξ^-	$\bar{\Xi}^+$	1322	1.6×10^{-10} s
		Omega	Ω^-	Ω^+	1672	0.8×10^{-10} s

[1]C. Amsler et al. (Particle Data Group), Physics Letters B667, 1 (2008) and 2009 partial update for the 2010 edition available on the PDG WWW pages (URL: http://pdg.lbl.gov/), unless otherwise noted

[2]Data from multiple sources

[3]Value depends on decay mode.

PHOTO CREDITS

Key: (t) top; (c) center; (b) bottom; (l) left; (r) right

COVER

Henrik Sorensen/Digital Vision via Getty Images

FRONT MATTER

vi kotoffei/Shutterstock.com; **viii** WLADIMIR BULGAR /SCIENCE PHOTO LIBRARY via Getty Images

CHAPTER 1

1 ZUMA Press, Inc./Alamy Stock Photo; **3** kalasek/iStock via Getty Images; **9** Phototribe/Shutterstock.com

CHAPTER 2

11 NASA; **12** leosapiens/Shutterstock.com; **19** ViktorKozlov /Shutterstock.com; **25** PKM1/E+ via Getty Images

CHAPTER 3

29 LEE SNIDER PHOTO IMAGES/Shutterstock.com; **34** Mark D Bailey/Shutterstock.com; **35** Artokoloro/Alamy Stock Photo

CHAPTER 4

41 Debby Wong/Shutterstock.com; **47** Robertus Pudyanto /Getty Images AsiaPac via Getty Images Sport

CHAPTER 5

53 Empato/E+ via Getty Images; **55** sciencepics/Shutter stock.com; **61** ©grifco/123RF.COM; **63** Sean Pavone/Shutter stock.com

CHAPTER 6

65 y_carfan/iStock Editorial via Getty Images; **66** Alexey Fursov/Shutterstock.com; **73** www.jetphotos.co.uk/Alamy Stock Photo

CHAPTER 7

79 Science History Images/Alamy Stock Photo

CHAPTER 8

91 V_E/Shutterstock.com; **93**br rangizzz/Shutterstock.com; **93**bl azure1/Shutterstock.com; **93**t Mark Herreid/Shutter stock.com; **95** Vitalii Nesterchuk/Shutterstock.com; **96** Scott Barbour/The Image Bank RF via Getty Images

CHAPTER 9

99 testing/Shutterstock.com; **101** Antagain/E+ via Getty Images; **103** Michael Krinke/E+ via Getty Images; **104** Flash Movie//iStockphoto via Getty Images; **110** sturti /iStockphoto via Getty Images

CHAPTER 10

111l Glasshouse Images/The Image Bank RF via Getty Images; **111**r Henry Steadman/Photodisc via Getty Images; **113** NASA Photo; **117** Ed-Ni-Photo/iStockphoto via Getty Images; **118** Stockbyte/Stockbyte via Getty Images; **119** Chris Hackett/Tetra images RF via Getty Images

CHAPTER 11

121 sandsun/iStockphoto via Getty Images; **124** kzen /Shutterstock.com; **129** ©NOLIMITPICTURES/iStockphoto via Getty Images; **130** Moof/Image Source via Getty Images; **135** choochart choochaikupt/iStockphoto via Getty Images

CHAPTER 12

143 F-15C Eagles from the 67th Fighter Squadron at Kadena Air Base, Japan/U.S. Air Force photo/Tech. Sgt. Angelique Perez/wikimedia commons/public domain; **149** image BROKER/Alamy Stock Photo; **151** Boyloso/Shutterstock.com

CHAPTER 13

153 Phil Leo/Michael Denora Images/Digital Vision via Getty Images; **154** Eivaisla/Shutterstock.com; **157** Arseniy45/iStockphoto via Getty Images

CHAPTER 14

159 "The Monitor and Merrimac: The First Fight Between Ironclads," a chromolithograph of the Battle of Hampton Roads, produced by Louis Prang & Co., Boston/1886/Library of Congress/Public Domain; **164** LP2 Studio/Shutterstock .com; **167** Martin Wahlborg/iStockphoto via Getty Images; **168** Worawut Prasuwan/iStockphoto via Getty Images

CHAPTER 15

171 HappyKids/E+ via Getty Images; **173** Bandersnatch /Shutterstock.com

CHAPTER 16

179 © Alfred Benjamin/SCIENCE SOURCE; **187** © GIPhoto Stock/SCIENCE SOURCE

CHAPTER 17

195 Chonticha Vatpongpee/EyeEm via Getty Images; **199** Inus Grobler/EyeEm via Getty Images; **201** © Andrew Lambert Photography/SCIENCE SOURCE; **206** pengpeng/E+ via Getty Images; **209** Don Bayley/E+ via Getty Images; **213** TheSnake19/iStockphoto via Getty Images

CHAPTER 18

217 Andrew Brookes AB Still Ltd/Image Source via Getty Images; **222** Wittayayut Seethong/EyeEm via Getty Images

CHAPTER 19

223 canadianPhotographer56/Shutterstock.com; **229** Toshe_O/iStockphoto via Getty Images; **233** Christian Petersen-Clausen/Moment Open via Getty Images

CHAPTER 20

235 KITTISAK PHONA/Shutterstock.com; **241** Caspar Ben son/fStop via Getty Images; **247** Jorge Villalba/iStockphoto via Getty Images

CHAPTER 21

255 Menna and Family Hunting in the Marshes/c.1422 - c.1411 BC/New Kingdom/Tomb of Menna, c.1422-1390 BC /modified/Public Domain; **265** PJF Military Collection/Alamy Stock Photo; **267** Richard A. Seeley/BJU Press

CHAPTER 22

273 Roberto/iStockphoto via Getty Images; **277** Laurent Hamels/PhotoAlto via Getty Images

CHAPTER 23

281 © Burrus/NIST/SCIENCE SOURCE; **284** strizh-/Shutter stock.com

CHAPTER 24

287 Thomas M. Barwick INC/Stone RF via Getty Images; **291** Krasowit/Shutterstock.com

CHAPTER 25

295 alefbet/Shutterstock.com; **297** D-Keine/E+ via Getty Images; **298** MOHD NOR HASEN BIN SUDIN/Shutterstock .com; **303** © CERN/SCIENCE SOURCE; **307** DE AGOSTINI PICTURE LIBRARY/De Agostini Editorial via Getty Images; **309** © Jose A. Bernat/Moment RF via Getty Images

BACK MATTER

317 Kali Nine LLC/E+ via Getty Images; **318** PhotoAlto/Alix Minde/ PhotoAlto via Getty Images; **319**t Science Photo Library/Science Photo Library RF via Getty Images; **319**b MShieldsPhotos/Alamy Stock Photo; **320** Sunyixun/MirageC /Moment RF via Getty Images; **321** MA8/Shutterstock .com; **322**t ChaosMaker/Shutterstock.com; **322**b Andrei Nekrassov/Shutterstock.com; **323**t © MARTYN F. CHILL MAID/SCIENCE SOURCE; **323**b Andrei Nekrassov/Shutter stock.com; **324–25** Phyphox screenshot/Creative Commons Attributions License; **326–27** Granted permission by Tracker via Doug Brown

PERIODIC TABLE
of the Elements

atomic number	86
radioactive	☢
name	Radon
symbol	Rn
atomic mass	(222)
electron structure by energy level	*2, 8, 18, 32, 18, 8*

atomic mass
- rounded to hundredths place
- mass number of isotope with longest known half-life indicated by ( )

	1	2	3		4	5	6	7	8	9
1	1 Hydrogen **H** 1.01 *1*									
2	3 Lithium **Li** 6.94 *2, 1*	4 Beryllium **Be** 9.01 *2, 2*								
3	11 Sodium **Na** 22.99 *2, 8, 1*	12 Magnesium **Mg** 24.31 *2, 8, 2*								
4	19 Potassium **K** 39.10 *2, 8, 8, 1*	20 Calcium **Ca** 40.08 *2, 8, 8, 2*	21 Scandium **Sc** 44.96 *2, 8, 9, 2*		22 Titanium **Ti** 47.87 *2, 8, 10, 2*	23 Vanadium **V** 50.94 *2, 8, 11, 2*	24 Chromium **Cr** 52.00 *2, 8, 13, 1*	25 Manganese **Mn** 54.94 *2, 8, 13, 2*	26 Iron **Fe** 55.85 *2, 8, 14, 2*	27 Cobalt **Co** 58.93 *2, 8, 15, 2*
5	37 Rubidium **Rb** 85.47 *2, 8, 18, 8, 1*	38 Strontium **Sr** 87.62 *2, 8, 18, 8, 2*	39 Yttrium **Y** 88.91 *2, 8, 18, 9, 2*		40 Zirconium **Zr** 91.22 *2, 8, 18, 10, 2*	41 Niobium **Nb** 92.91 *2, 8, 18, 12, 1*	42 Molybdenum **Mo** 95.95 *2.8, 18, 13, 1*	43 ☢ Technetium **Tc** (98) *2, 8, 18, 13, 2*	44 Ruthenium **Ru** 101.07 *2, 8, 18, 15, 1*	45 Rhodium **Rh** 102.91 *2, 8, 18, 16, 1*
6	55 Cesium **Cs** 132.91 *2, 8, 18, 18, 8, 1*	56 Barium **Ba** 137.33 *2, 8, 18, 18, 8, 2*	57 Lanthanum **La** 138.91 *2, 8, 18, 18, 9, 2*	*	72 Hafnium **Hf** 178.49 *2, 8, 18, 32, 10, 2*	73 Tantalum **Ta** 180.95 *2, 8, 18, 32, 11, 2*	74 Tungsten **W** 183.84 *2, 8, 18, 32, 12, 2*	75 Rhenium **Re** 186.21 *2, 8, 18, 32, 13, 2*	76 Osmium **Os** 190.23 *2, 8, 18, 32, 14, 2*	77 Iridium **Ir** 192.22 *2, 8, 18, 32, 15, 2*
7	87 ☢ Francium **Fr** (223) *2, 8, 18, 32, 18, 8, 1*	88 ☢ Radium **Ra** (226) *2, 8, 18, 32, 18, 8, 2*	89 ☢ Actinium **Ac** (227) *2, 8, 18, 32, 18, 9, 2*	**	104 ☢ Rutherfordium **Rf** (267)	105 ☢ Dubnium **Db** (268)	106 ☢ Seaborgium **Sg** (269)	107 ☢ Bohrium **Bh** (270)	108 ☢ Hassium **Hs** (269)	109 ☢ Meitnerium **Mt** (278)

*	58 Cerium **Ce** 140.12 *2, 8, 18, 19, 9, 2*	59 Praseodymium **Pr** 140.91 *2, 8, 18, 21, 8, 2*	60 Neodymium **Nd** 144.24 *2, 8, 18, 22, 8, 2*	61 ☢ Promethium **Pm** (145) *2, 8, 18, 23, 8, 2*	62 Samarium **Sm** 150.36 *2, 8, 18, 24, 8, 2*
**	90 ☢ Thorium **Th** 232.04 *2, 8, 18, 32, 18, 10, 2*	91 ☢ Protactinium **Pa** 231.04 *2, 8, 18, 32, 20, 9, 2*	92 ☢ Uranium **U** 238.03 *2, 8, 18, 32, 21, 9, 2*	93 ☢ Neptunium **Np** (237) *2, 8, 18, 32, 22, 9, 2*	94 ☢ Plutonium **Pu** (244) *2, 8, 18, 32, 24, 8, 2*

- Alkali metals
- Alkaline-earth metals
- Transition metals
- Post-transition metals
- Metalloids
- Inner transition metals
- Nonmetals
- Halogens (also nonmetals)
- Noble gases
- ☢ Radioactive isotopes

10	11	12	13	14	15	16	17	18
								2 Helium **He** 4.00 *2*
			5 Boron **B** 10.81 *2, 3*	6 Carbon **C** 12.01 *2, 4*	7 Nitrogen **N** 14.01 *2, 5*	8 Oxygen **O** 16.00 *2, 6*	9 Fluorine **F** 19.00 *2, 7*	10 Neon **Ne** 20.18 *2, 8*
			13 Aluminum **Al** 26.98 *2, 8, 3*	14 Silicon **Si** 28.09 *2, 8, 4*	15 Phosphorus **P** 30.97 *2, 8, 5*	16 Sulfur **S** 32.06 *2, 8, 6*	17 Chlorine **Cl** 35.45 *2, 8, 7*	18 Argon **Ar** 39.95 *2, 8, 8*
28 Nickel **Ni** 58.69 *2, 8, 17, 1*	29 Copper **Cu** 63.55 *2, 8, 18, 1*	30 Zinc **Zn** 65.38 *2, 8, 18, 2*	31 Gallium **Ga** 69.72 *2, 8, 18, 3*	32 Germanium **Ge** 72.63 *2, 8, 18, 4*	33 Arsenic **As** 74.92 *2, 8, 18, 5*	34 Selenium **Se** 78.97 *2, 8, 18, 6*	35 Bromine **Br** 79.90 *2, 8, 18, 7*	36 Krypton **Kr** 83.80 *2, 8, 18, 8*
46 Palladium **Pd** 106.42 *2, 8, 18, 18*	47 Silver **Ag** 107.87 *2, 8, 18, 18, 1*	48 Cadmium **Cd** 112.41 *2, 8, 18, 18, 2*	49 Indium **In** 114.82 *2, 8, 18, 18, 3*	50 Tin **Sn** 118.71 *2, 8, 18, 18, 4*	51 Antimony **Sb** 121.76 *2, 8, 18, 18, 5*	52 Tellurium **Te** 127.60 *2, 8, 18, 18, 6*	53 Iodine **I** 126.90 *2, 8, 18, 18, 7*	54 Xenon **Xe** 131.29 *2, 8, 18, 18, 8*
78 Platinum **Pt** 195.08 *2, 8, 18, 32, 17, 1*	79 Gold **Au** 196.97 *2, 8, 18, 32, 18, 1*	80 Mercury **Hg** 200.59 *2, 8, 18, 32, 18, 2*	81 Thallium **Tl** 204.38 *2, 8, 18, 32, 18, 3*	82 Lead **Pb** 207.24 *2, 8, 18, 32, 18, 4*	83 Bismuth **Bi** 208.98 *2, 8, 18, 32, 18, 5*	84 Polonium **Po** (209) *2, 8, 18, 32, 18, 6*	85 Astatine **At** (210) *2, 8, 18, 32, 18, 7*	86 Radon **Rn** (222) *2, 8, 18, 32, 18, 8*
110 Darmstadtium **Ds** (281)	111 Roentgenium **Rg** (282)	112 Copernicium **Cn** (285)	113 Nihonium **Nh** (286)	114 Flerovium **Fl** (289)	115 Moscovium **Mc** (290)	116 Livermorium **Lv** (293)	117 Tennessine **Ts** (294)	118 Oganesson **Og** (294)

63 Europium **Eu** 151.96 *2, 8, 18, 25, 8, 2*	64 Gadolinium **Gd** 157.25 *2, 8, 18, 25, 9, 2*	65 Terbium **Tb** 158.93 *2, 8, 18, 27, 8, 2*	66 Dysprosium **Dy** 162.50 *2, 8, 18, 28, 8, 2*	67 Holmium **Ho** 164.93 *2, 8, 18, 29, 8, 2*	68 Erbium **Er** 167.26 *2, 8, 18, 30, 8, 2*	69 Thulium **Tm** 168.93 *2, 8, 18, 31, 8, 2*	70 Ytterbium **Yb** 173.05 *2, 8, 18, 32, 8, 2*	71 Lutetium **Lu** 174.97 *2, 8, 18, 32, 9, 2*
95 Americium **Am** (243) *2, 8, 18, 32, 25, 8, 2*	96 Curium **Cm** (247) *2, 8, 18, 32, 25, 9, 2*	97 Berkelium **Bk** (247) *2, 8, 18, 32, 27, 8, 2*	98 Californium **Cf** (251) *2, 8, 18, 32, 28, 8, 2*	99 Einsteinium **Es** (252) *2, 8, 18, 32, 29, 8, 2*	100 Fermium **Fm** (257) *2, 8, 18, 32, 30, 8, 2*	101 Mendelevium **Md** (258) *2, 8, 18, 32, 31, 8, 2*	102 Nobelium **No** (259) *2, 8, 18, 32, 32, 8, 2*	103 Lawrencium **Lr** (266) *2, 8, 18, 32, 32, 8, 3*